Language and Globalization

Series Editors

In the context of current political and social developments, where the national group is not so clearly defined and delineated, the state language not so clearly dominant in every domain, and cross-border flows and transfers affects more than a small elite, new patterns of language use will develop. This series aims to provide a framework for reporting on and analysing the lingustic outcomes of globalization and localization.

Sari Pietikäinen

Cold Rush

Critical Assemblage Analysis of a Heating Arctic

palgrave
macmillan

Sari Pietikäinen
Department of Language
and Communication Studies
University of Jyväskylä
Jyväskylä, Finland

ISSN 2947-7506 ISSN 2947-7514 (electronic)
Language and Globalization
ISBN 978-3-031-63994-4 ISBN 978-3-031-63995-1 (eBook)
https://doi.org/10.1007/978-3-031-63995-1

This Palgrave Macmillan imprint is published by the registered company Springer Nature Switzerland AG
The registered company address is: Gewerbestrasse 11, 6330 Cham, Switzerland

Acknowledgements

The story of this book has unfolded over a long period of time, which is one way to say that it has taken me a long time to think through and write this book. The scope of the book, together with the research project it draws on and the time it has taken to write this book, means that there are many individuals and institutions I would like to thank.

I begin by thanking all of the participants in the Cold Rush research project for sharing their experiences and practices with me. My heart-felt gratitude goes to the key participants, called in this book as Heli, Leena, Lucy, Matti, Pentti, Simo, and Taimi, for their generosity and patience in discussing Cold Rush issues and reading the earlier versions of this book.

The project and writing of this book would not have been possible without institutional support. I warmly thank the Academy of Finland, the Association of Finnish Non-Fiction Writers, the Department of Language and Communication Studies. My appreciation goes to the students and colleagues in the department, especially Mika Lähteenmäki and the Discourse Hub crew: Elena Barrett, Sigurd D'hondt, Ari Kuismin, Riikka Nissi, and Pauliina Puranen. Thank you for productive conversations that have enriched this book in countless ways.

Over the years, I have discussed various ideas and arguments related to this book in conferences, seminars, and workshops, and I want to thank all participants for their valuable insights and critical questions that have helped me develop the book further. This book has benefitted from insightful discussions and intellectual exchange with several scholars, and my sincere thanks go to Lindsay Bell, Luke Cantarella, Jose A. Corter-Vazques, Kellie Gonçalves, Florian Hiss, Patricia LaMarre, Gavin Lamb, Christina Hegel, Erez Levon, Marie Maegaard, Luisa Martín-Rojo, Tommaso Milani, Andreas Nuottaniemi, Charlotta Seiler Brylla, Hilde Sollid, Crispin Thurlow, Laura Villa, Virginia Zavala Cisneros and Henning Årman. My time as an ARC visiting professor at the City University of New York was instrumental in developing this book, and I am grateful to my host, Professor Ofelia García, colleagues, and students at CUNY for the most inspiring conversations.

The excellent Cold Rush project team has provided me with a valuable terrain to examine and discuss ideas discussed in this book, and I would like to thank Kori Allan, Alexandre Duchêne, Monica Heller, Anna-Liisa Ojala, and Maiju Strömmer for inspiring collaboration and dedicated teamwork. Alexandre and Monica were instrumental for the productive development of the project and invaluable partners in thinking through the Cold Rush and it's connections within and beyond the Arctic. Alexandre had a vital role in creating the Cold Rush Science Cards as our way of communicating the key findings. The advisory board members, Niko Besnier, Susana Narotzky, Juan Pujolar, and Monica Tennberg, provided valuable feedback and support at critical points in the project.

The writing of this book has benefitted significantly from the help and support of three excellent scholars and writers. Anne Mäntynen is a friend and scholar with whom everything feels plausible, and our collaborations and conversations have made writing this book much lighter than it would have been otherwise. Helen Kelly-Holmes and I have worked together on several projects in the Arctic and beyond, and it is always hugely productive and great fun to work with her. This is also true for this book, as Helen has generously provided me with detailed and helpful suggestions that have greatly improved the clarity and coherence of this book. Ingrid De Saint-Georges has been an unwavering support and source of inspiration throughout the writing process of this book. Our collective efforts to enjoy and experiment with writing has helped me tremendously, and Ingrid's curious mind and bold attitude have energized my writing. Ingrid was the first to see a sense in this book and the first reader of the various versions of this book. I doubt I would have finished it without her.

This book would not exist without my family, Pekka, Helmi, Aapo, and Hilla. Pekka was the reason I first came to the Arctic and why I am still there some 30 years later. He was instrumental in creating the Cold Rush photo archive, which consists of thousands of photos that we took and curated together. He has also provided invaluable help and support for the photographs featured in this book. Hilla's input for the Cold Rush photo archive and finalizing the manuscript has been invaluable, Helmi has been proofreading the text with impossible deadlines and Aapo has been an unweavering source for reflections. Our children have grown into young adults over the years it has taken me to write this book, and I have been amazed and inspired by their critical and insightful views about the Cold Rush and climate change that they have witnessed firsthand since their childhood. They have helped me to understand the gravity of the implications of climate change in the Arctic, provided me with support and chocolate, and patiently listened to my rants of the week, most often about Deleuze.

Contents

List of Figures

1

Arctic Cold Rush

1.1 Accelerating Cold Rush

Fig. 1.1 The ice-braided Tornio River

S. Pietikäinen, *Cold Rush*, Language and Globalization,
https://doi.org/10.1007/978-3-031-63995-1_1

Every June, the wild Atlantic salmon returns to its natal Arctic river, Tornionjoki, to spawn (Fig. 1.1). This over 500-kilometre-long border river between Sweden and Finland runs free from power plants, making it Europe's longest free-flowing river for salmon. As the salmon begin their run to the upper reaches of the river, so too do the fishers. The fishing fever spreads with the quickening pace of the salmon run. Small villages alongside the riverbanks, forgotten during the long Arctic winter, become buzzing hubs for thousands of fishers wanting to try their luck with the king of the fish. Handwritten signs, typically written in Finnish but at times in English, too, pop up along roads, advertising cottages, boats, and guides for rent, as the locals, mainly seniors with working histories in the local mines, forestry, and tourism business, try to earn some extra money in this fever pitch. More entrepreneurial locals with tourism businesses in the nearby ski resorts have purchased old village schools, emptied of pupils already in the 1970s by the big recession, and turned them into fishing tourism sites, often with support from various European Union funding schemes to keep the peripheries vibrant.

Fishing is a speculative business. How to catch a salmon remains a mystery, and the locals say that salmon is given not caught. As the fishing season progresses, speculations run high on when, where, and how to catch the fish. Some use a smartphone application to calculate the ideal time for fishing. Others trust indexes of weather, while the majority simply fish around the clock, seven days a week. The lucky ones who catch fish get their name, the name of the lure, and the size of the fish written down on a large notice board hanging on an outer wall of the old school building, a popular destination for locals and fishing fans for an evening stroll. The same information is posted on a website showing the fishing statistics of past summers, too. As summer progresses, a pattern emerges: some catch more fish than others. Envy grows, together with speculations about reasons for the success of some and the failure of many.

In the last few summers, many salmon have had injuries, a type of rash, on their sides and a few fish have been found dead in the river. There is much speculation about this: maybe climate change is turning the river too warm for salmon. Alternatively, the Baltic Sea, to which the river connects, has become too polluted and shallow. Or perhaps it is pollution from iron ore mines at the river's upper reach. Some people eat fish now with some hesitation, and a few have stopped altogether.

The best course of action for the future of the area is hotly debated and full of contradictions. Fishing tourism has lots of potential, but mining is valued as an all-a-year-round job. Tourism is a relatively recent source of income in the region, where mining has over 400-year history. Green, sustainable practices often brought in by outsiders from the South are pitted against the traditional livelihoods in forestry, farming, and fishing. Fishing fever is intertwined with the Arctic's mining and tourism rushes in the region already struggling with ageing and a declining population, a fading service infrastructure, and reducing tax revenue. Visions about how to cope, manage, and profit from the dynamics of hope and harm of the rush are diverse, incomplete, and debated. What is clear, however, is that as the rush runs high in the Arctic, the stakes for individuals, communities, and the environment rise.

This book is about the Cold Rush: an intensified race for extraction and protection of Arctic natural resources fuelled by accelerating climate change, expanding markets, and a novel revaluating of commons as commodities. The rush turns mining into a good job, wild berries into superberries, and a car ride in an Arctic night into a Northern Lights chase experience. Reminiscent of the Gold Rush in the nineteenth and twentieth centuries, the Arctic Cold Rush creates novel opportunities, high ecological risks, and ongoing political debates about how to best govern, extract, and protect natural resources. The Cold Rush increases the region's strategic importance while opening it up to new ideological confrontations over the value of natural resources embedded in power dynamics.

The sense of rush is heightened by accelerating climate change in the Arctic. The Arctic is warming four times faster than the rest of the world, with tangible and unprecedented consequences (Li et al. 2023; Rantanen et al. 2022). The Ice Sea is melting in front of us, prompting speculations about the risks and the gains of the change. Shrinking sea ice endangers the lives of many species, such as polar bears, reindeer, and certain fish, while creating new opportunities for southern species. Similarly, many traditional livelihoods, like reindeer herding, are becoming endangered while new economic opportunities for oil drilling, mining, and logistics are emerging. Significantly shorter trans-arctic shipping routes hold enormous potential for time and cost savings. The Arctic region has generated an economic buzz as its vast, partly still uncharted natural resources become more accessible due to climate change (Larsen and Huskey 2015). At the same time, a wide range of regional governments, indigenous nations, and NGOs are pushing for more binding and stricter environmental laws and policies for the Arctic. The rush to extract and protect Arctic resources is simultaneous.

As opportunities for exploiting Arctic natural resources emerge, so too do new possibilities for the commodification of newly valued commons, such as snow, water, peripherality, and silence. Branding clean air, packaging Santa Claus, and capitalising on pure water are all emerging economic strategies to profit from ongoing, interconnected political, ecological and economic changes. The Cold Rush is accelerated in its most speculative forms in this land of superlatives. It has Europe's cleanest air and the purest water, which can be drunk directly from a well. In summer, the sun does not set at all and the white nights create exceptional growing conditions for wild berries, famous for their high nutritional density. Every winter, during the Polar Nights season, the sun does not rise at all, creating the perfect scenery for a tourist winter wonderland. Snow is guaranteed every winter, an opportunity utilized by global sports events and tourism. The region's geographical Aurora Borealis zone makes sightings of the Northern Lights a regular event.

The growth of resource extraction, transportation infrastructure, and tourism industry in the region has led to an influx of economic and human capital on the one hand and tension between the ecological gains and risks on the other. It is fuelled by an enduring colonialist perception of the region as a last resource frontier, an empty land, waiting to be grabbed, developed, and utilized (Spangen et al. 2015; Junka-Aikio 2023; Lehtola 2002). It has also brought forth struggles and strategies for coping with and profiting from the rush and the inevitable cycles of economic boom and bust. These changes also reveal processes of flexible and transnational capitalism, amplified by rapidly progressing climate change. Also, the recent geopolitical changes and tensions have increased the political significance of the Finnish Arctic. From this perspective, one outcome of the escalation of the Russia-Ukraine war in 2022 was that Finland, which had valued its military neutrality since 1945, joined the North Atlantic military alliance (NATO) in 2023, making Finland's over 1000 km long Eastern border the border between Western allies and Russia. Another outcome was an energy crisis, as Finland, like many other European countries, relied heavily on Russian gas. The urgent need for alternative energy production resources has increased further the value of Arctic natural resources.

These transformations are highly consequential. Arctic natural resources undergo intertwined ecological, political, economic, and social transformations, imbuing them with varying values (Carrasco et al. 2023; Paajanen 2024). As the value of Arctic natural resources changes, so too do relationships between resources, people, and governance. These changes impact not only how people make a living but also how their trajectories, capacities, and roots are valued or disregarded as well as their ability to cope with and reshape these changes.

In this book, I examine this ambivalent, emerging, and changing conjuncture of the Cold Rush in the Finnish Arctic, in the region called Lapland. This northernmost reach of the planet has become a high-stakes ecological, economic, and political hotspot—a laboratory for what is happening globally in the Polar Arctic region. It is a heterogeneous space understood simultaneously as indigenous land within indigenous Sámi politics, the last frontier from a colonial perspective, and a periphery under the modernist Finnish nation-state regime.

This heterogeneity makes the political economy of the Finnish Arctic complex. The region is part of indigenous Sámiland, which has a long colonial history of being constructed as a frontier. The Sámi history has gradually been overwritten by the histories of southern kings, nation-states, and their colonial practices. The Crown of Sweden ruled the area until 1808, when the Russian Tsar took it over, ruling until Finland gained independence in 1917. For a century, the State of Finland has treated the region as a periphery, a resource reservoir for the benefit of the south of the country where the capital and the majority of the population are located. Today, the northernmost tip of the Finnish Arctic region belongs to the legally recognized indigenous Sámi homeland, *Sápmi*. While the cultural rights of the indigenous Sámi people in Finland were legally recognized in 1991, the legal and political struggle for land rights has continued over decades (Valkonen 2018, 2009). Currently, the Sámi do not have special rights to the Arctic land in Finland. The State of Finland has not signed the *Indigenous and Tribal Peoples Convention*, 1989, No. 169 convention, which requires settlement of the land rights in the Arctic. However, most Arctic land continues to be legally subscribed under the state land category and governed by the State (Kuokkanen 2023; Lehtola 2002; Valkonen et al. 2022).

For the past hundred years that the State of Finland has governed the region, its economy has relied on forestry, mining, and tourism. The mining industry has a 400-hundred-year history in the region, intertwined with settler colonialism and the emergence of capitalism, and expeditions continue to this today (see Chapter 3). Finland is a top destination for global mining investors because the land is mineral-rich, and the country has fewer mining restrictions than many other European countries (Kröger 2016; Lassila 2018). The social value of mining in the Finnish Arctic is constructed through permanent jobs and solid tax revenue, both valued in a region with a long history of unemployment and poverty (Pietikäinen and Allan 2021). At the same time, mining is controversial as it endangers nature and, by extension, nature tourism, the other primary industry in the area.

The tourism industry cultivates the idea of the Arctic as the last frontier, a land for discovering the magic of Lapland: Northern Lights, white nights, pristine snow, clear water, and pure air, and the gifts of nature such as berries, mushrooms, fish, and reindeer (see Chapters 4 and 5). While these two constructions of the Arctic—on the one hand, a busy, technologically driven, mining-friendly hotspot and, on the other, pure, calm, and untouched nature—are contradictory, these frictions are ironed out by the longstanding discourse of the region as a place where there is room for everything (Jokinen 2019). This echoes the colonial understanding of the region as empty terra nullius. Political decisions about resource extraction employ discourses of progress and modernization, touting these economic activities as bringing wealth and well-being to the periphery (Bebbington et al. 2014). The political and economic control over Arctic natural resources is unsurprisingly an object of heated debate, juxtapositions, and paradoxes.

Currently, the Finnish Arctic is simultaneously undergoing two massive effects of modernity: the acceleration of climate change and the revaluing of its natural resources. Neither process is linear nor complete; rather, they are contradictory and open-ended. This moment is full of complex contradictions, opportunities, and risks. Resource extraction for minerals, as well as tourism, transportation, and trade infrastructure, provide some economic security, yet simultaneously, it threatens the well-being of the environment, communities, and people dependent on the very same natural resources. These changes provoke economic, cultural, and political restructuring of natural resources and relationships around them, involving various actors, including indigenous nations, local communities, nation-state regimes, and supranational investors. Yet these heating Arctic hotspots reside next to "cooling", almost forgotten places where fewer people live, and, year by year, more shops, schools, and other services are closed down. The properties have lost value, and the once-thriving mining and timber towns have become ghost towns. Utopia and dystopia, prosperity and poverty co-exist and intertwine.

The Cold Rush gets its fuel from this terrain of possibilities and uncertainties, and, in this book, I focus on how people struggle, strategize, and profit from this ongoing, complex, and multidirectional change. Examining this requires an approach that opens up and keeps up with the reasons and rhizomes of these transformations and the restructuration of the relationships around them. In this book, I develop and apply a theoretical approach to analyse and explain the multiplicity of these processes, their conditions, and their consequences in this book. This approach, which I call *critical assemblage analysis* (see Chapter 2), builds on my previous work at the intersection of critical discourse analysis, critical sociolinguistics, and ethnography (see, e.g., Pietikäinen 2016; Heller et al. 2018; Pietikäinen et al. 2016; Pietikäinen and Mäntynen 2019) but expands towards works by philosophers Michel Foucault (1926–1984) and Gilles Deleuze (1925–1995) including his collaboration with a psychoanalyst Felix Guattari (1930–1992).

1.2 Connecting Flows: Foucault and Deleuze in the Arctic

Fig. 1.2 Unfolding snow

As I gradually unravelled the multiple, overlapping, and changing processes of revaluing Arctic natural resources, I realized that I need thinking tools that help me grasp and make sense of these emerging complexities and network of connections. I wanted to understand the historical and current conditions that seem to accelerate the Cold Rush and explore both enduring and ephemeral revaluing processes of natural resources. I was curious to notice how the same elements, like employment, workers, progress, and the vastness of the Arctic nature, seemed to reappear in different places and at different times, whereas some other elements, such as health, sustainability, and colonialism, appeared only in fleeting moments or specific places. Also, some of the revaluing attempts seemed to work out as planned, but not all of them, and a few caused friction and debates. I was also puzzled by the ways the revaluations reconfigured relationships between people but also between land, access, categories, time, and prospects—between all kinds of "stuff". I was looking for tools to map these processes, entangelments and reconfigurations, analyse their content and expressions, and study the logics that organize them and the connections between them. I turned to two key figures in discourse studies, Foucault and Deleuze. I have been reading and working with Foucault's ideas, particularly those related to *discourse*, power, and knowledge, since the start of my studies, and the works by Deleuze and Deleuze and Guattari related to *rhizome* and *assemblage* became relevant over ten years ago when I was trying to understand connections between places and processes that I have been studying in the Arctic. There certainly are many other relevant and newer theories and concepts, such as posthumanism (Braidotti 2019) anthropocene (Tsing 2015) new materialism (Bennett 2010) and post-colonialism (Bignall and Patton 2010) for examining the Cold Rush (see Chapter 2), and many of them engage with the works by Foucault, Deleuze, and Guattari. In my case, thinking with, through and towards the concepts of discourse, assemblage and rhizome developed by these scholars helped me to examine and explain what is at stake in the various revaluing processes of Arctic natural resources (Fig. 1.2).

The biographies of these two prominent French philosophers are intertwined, and their works resonate with each other. Michel Foucault and Gilles Deleuze are central figures in post-War post-structuralist French philosophy with enduring and emerging impacts in multiple fields beyond philosophy. They both worked in Paris at the same time, from the 1960s until Foucault's death in 1984. They shared the same philosophical heritage, interest in developing Nietzsche's thought, and some political activism, including the May 1968 protests and the Prison Information Group, created by Foucault in the early 1970s. Despite their many philosophical, political, and biographical connections, they did not actually work together much while they were in conversation with each other, and their work intersects in multiple ways (Morar et al. 2016: 1–2). Deleuze (1986/1988) described these connections as having many "parallels" between their work, and Foucault (1969/1972) famously stated that this century would become known as "Deleuzian". Towards the end of the 1970s and 1980s, they became more distant due to different political and philosophical interests and disagreements, but they kept reading each other's work. At Foucault's funeral in 1984, Deleuze read an excerpt from the introduction to Foucault's recently-published book *The Use of Pleasure* (1984). In the following academic year (1985–1986), Deleuze dedicated a lecture series to the work of Foucault at the University of Paris VIII (1985–1986) and, a few years later, published his interpretation of the particular logic of Foucault and shifts in his thinking in the book titled *Foucault* (1986/1988) (Dosse 2016). In the current book, I approach some aspects of their work through the lens of the Cold Rush, and I highlight selected connections, entangelments, and differences relevant to my examination of revaluing Arctic natural resources (for further discussion on the relationship between Deleuze and Foucault, see, e.g., Dosse 2010; Morar et al. 2016).

Both Foucault and Deleuze were sharp critics of the structuralist approach to social and political organization, and they were united in rejecting any essentialist idea. They shared an interest in intellectuals' roles and power, as well as critical social, political, and economic processes and practices, while their methods and motivations for approaching social issues and processes differed (Morar et al. 2016: 3). Foucault, whose one key interest was truth, engaged with archaeological and genealogical analysis of knowledge, whereas Deleuze, with his longtime collaborator, psychoanalyst Felix Guattari, was interested in desire as a force in the capitalist system, and developed *schizoanalysis*. Foucault's emphasis on historical development prioritizes textual data, while Deleuze's focus on the interaction between virtual and actual becomings encompasses discursive, material, and affective dimensions. Foucault's work can be characterized as being driven by an interest in the dynamics of power and knowledge through discourse and the impact of subjectivity, while Deleuze's work engaged with concepts such as assemblage, rhizome, and desire to address the multiplicity of the world challenging the binary way of thinking and existing and embracing creativity thinking (Dosse 2010: 306–30; Morar et al. 2016: 1–2).

Besides the different methods of analysis, they were also driven by different motivations: Foucault wanted to understand the emergence of liberalism, while Deleuze and Guattari sought to understand the schizophrenic breakdown of contemporary capitalism. Their approaches also shaped the alternatives they proposed: Foucault's ethical self-transformation and Deleuze's revolutionary nomadism. In their comments on the differences in their work, Foucault was critical of the concept of desire, while Deleuze was unconvinced by the purpose of the concept of the subject and what he saw as an over-determinacy of power and the primacy of enunciation (Morar et al. 2016).

As I gradually map cartography around these two giants and their gigantic terms, I think of the Cold Rush as a complex and emergent interaction between material, discursive, and affective dimensions embedded in power relations under capitalist conditions. I believe the Cold Rush cannot be productively thought of without any of these dimensions, but they are not easy to think of together. The Cold Rush is not reducible to one thing or a single concept, but it is simultaneously material, discursive, and embodied; modernist and post-modernist; national and indigenous, global and local. It reflects a complex sense of belonging, access, and mobility. It is a historical but also an ongoing, emerging, unfinished, and speculative process. This makes the Cold Rush both exciting and challenging to research, pushing me to rethink how to examine the revaluing of Arctic natural resources (see Chapter 2).

In this book, I selectively zig-zag between different yet related concepts from Foucault and from Deleuze and Guattari. The concepts of *discourse*, *assemblage*, and *rhizome* are core concepts in the book, but I also engage with concepts of *regime*, *power*, *affect*, *de/reterritorialision*, *smooth* and *stratified spaces*, *desire*, *unfolding*, *code*, and *machine*. These concepts have become relevant as I examine the unfolding Cold Rush in the Finnish Arctic. Following Deleuzian thinking, these concepts overlap and have porous boundaries and rhizomatic connections, but they gain their specific function in a particular situation, in a particular moment of studying the Cold Rush (see more in Chapter 2). Written in a dialogue with Foucault (Deleuze and Foucault 1977: 208), Deleuze famously compares theory to a toolbox, suggesting that concepts are tools for analysis and critical and creative thinking. Against this backdrop, I explore and try out these concepts with my data, writing and thinking as I explore the gradually unfolding Cold Rush in the Finnish Arctic.

Deleuze uses the medial term palpation to describe scientific discovery's critical and creative process. Palpation is the physical examination of a lesion where the doctor senses it without directly experiencing it. Similarly, concepts seek to palpate something that cannot be directly comprehended (May 2005: 20–23). As I navigate and palpate this terrain of concepts, I map and create a conceptual path for examining the Cold Rush and potentially offering an alternative to examining complex, ongoing, unfinished processes beyond the context of Cold Rush. Deleuze also writes that understanding an author means being able to dream about them. It is not that much of a matter of going back and digging up an author's thoughts but trying to navigate an author by following the current of thought (Gros and Bankston 2016: 129). This is what I try to do: to follow the current of thoughts brought forth by these concepts and interweave them with my own thinking while examining the Cold Rush. In this chapter, I will briefly discuss the key terms—assemblage, rhizome, and discourse. This paves the way for a more in-depth discussion of the concepts and thinking of Foucault, and Deleuze and Guattari as well as critical assemblage analysis in Chapter 2.

Let's start with assemblage and rhizome. Think about improvised jazz music: you can be attentive to one instrument or melody, but only when you listen to how all instruments, rhythms, and melodies come together, do you get a full sense of the piece of music. That is jazz assemblage in action. Moreover, unlike the written piece of music that can be repeated almost identically over and over again, improvised jazz assemblage changes as conditions or function changes (see Tsing 2015: 158). That is the difference between rhizome and tree thinking. Taking another example from team sports like ice hockey: we can be fans of a particular player and follow their career and performance on the ice, but the game is always lost or won as a team: a successful team requires collaboration between its different elements. The team drills different game strategies on ice, but in the actual game, the strategies are adopted, abandoned, and changed as conditions change (Pietikäinen and Hegel 2021), this is the hockey assemblage on ice.

Deleuze and Guattari (1987) use assemblage to refer to a multi-temporal, heterogeneous arrangement of discourses, materialities, bodies, and affects powered by desire and influenced by capitalist transformations. The term is used to signal a provocative critique of hierarchical organization models and modernity's discourses and institutions, providing a promising starting point for thinking about the complexity, connectivity, and intersectionality of the Cold Rush elements and processes. Assemblage can be seen as a theoretical metaphor for an interconnected and irreducible multiplicity of ongoing processes. It also provides a strong critique of capitalism (see Chapter 2).

The key benefit to using assemblage to examine the Cold Rush is that it does not flatten the complex historical and emergent relationships between the heterogeneous elements that constitute the value of natural resources, nor does it disregard how they are intertwined with various historical as well as ongoing political, economic, and personal processes. Furthermore, it does not give a prior weight to any particular element, for example, discourse or language. Rather, examining the Cold Rush aims to explore how "things", be they material, discursive, or affective, come together to produce a particular value to the natural resources in a particular situation. What a particular assemblage consists of and how it is organized becomes an empirical question. Assemblage begins and ends with heterogeneity, and this helps the researcher to examine how, when, and with what kind of impact certain elements and relationships construct a situated understanding of the value of Arctic natural resources. Revaluing Arctic resources cannot be reduced to single stories or categories, and assemblage facilitates a more in-depth examination of the multiplicities and complexities inherent in value formations, as well as their political and critical underpinnings.

Assemblages are more than just a messy, complex collection of various things; they also consist of multiple lines that connect the elements and enable them to interact and produce together the assemblage's function. The concept of *rhizome* refers to these multiple lines and connections. Borrowed from botany, rhizome refers to nonhierarchical and decentred thinking with multiple entry points and without one hierarchical order or organising principle. Rhizomatic thinking opposes what Deleuze and Guattari describe as tree thinking, which has a prior organization, hierarchy or basic structure (Clarke and Parsons 2013). One of their well-known publications, *A Thousand Plateaus: Capitalism and Schizophrenia* (Deleuze and Guattari 1980/1987), is rhizomatically organized, meaning the chapters can be read separately. Like assemblage, rhizome does not assume a prior, universal logic, core, or structure across time and place, nor does it consider development or progression to be linear.

To examine further the rationalities, categories, and historical development of values related to Arctic natural resources and their economic transformations, I draw on Foucauldian understanding of the intertwined relationship between power, knowledge, and discourse in seeing discourses as a significant force in shaping our concepts, knowledge, and understandings of events, people, and processes (see Chapter 2). Foucault, echoing Deleuze's and Guattari's concept of assemblage, emphasized that discourse always has material and affective conditions and effects (Foucault 1977/1984). The Foucauldian understanding of discourse as a nexus of power, knowledge, and meanings which encode rationalities is critical for examining the consequences of how we talk about and see the value of Arctic natural resources. Foucault emphasizes the genealogy of concepts and categories and their historical evolution over time. As Foucault reminds us, power and knowledge are intertwined: As our understanding is intertwined with the concepts that we use, these very same concepts enable and limit the ways we can understand and speak about processes, events, and people (Foucault 1969/1972). Through the lens of discourse, I examine how discourses and the rationalities imbued in them are used to create, challenge, justify, and transcend their value. My particular focus is on the power of discourse in constructing various meanings, categories, and rationalities of these transformations, which connects this book to previous work in critical discourse studies, critical sociolinguistics, and linguistic anthropology about the role of discourse in value-making under changing economic, political, and cultural conditions of global capitalism.

Foucault's work has helped me examine the historicity of the processes while keeping an eye on the dynamics between power and knowledge, as well as the role of discourse in this dynamic. At the same time, work by Deleuze, and Deleuze and Guattari forces me to account for relationships and connections across different multiplicities and processes while avoiding binary thinking. Their capitalism critique (Deleuze and Guattari 1972/1977) helps me in situating these connections within the Arctic's historical and emerging capitalist conditions.

Assemblage resonates closely with Foucault's concept of apparatus (dispositif), which refers to the arrangement of relations in a heterogeneous system (Foucault 1972/1977: 194). Both assemblage and apparatus aim to replace structuralist concepts of organization by focusing on relationships and connections between elements in a heterogeneous system (Morar et al. 2016: 3). Legg (2011) suggests that assemblage and apparatus can be thought of dialogically since apparatus tries to capture the more enduring order between the relationship whereas assemblage emphasizes the emergent and changing nature of these relationships. In this book, I try to do exactly this by including the idea of apparatus as a more enduring order with the idea of emergent relationships in an assemblage. Moreover, by bringing together a Foucauldian understanding of discourse and apparatus, and a Deleuze-Guattarian approach to assemblage and rhizome, I attempt to include the idea of historical embeddedness of the assemblage elements, the political alignment of the relationships with power dynamics, and the endurance with ephemeral and emergent assemblage forces. This is what it means to me to think of Cold Rush with discourse and assemblage together.

The concepts of assemblage, rhizome, and discourse all underline the importance of investigating the interaction between discursive and material aspects of the revaluing of natural resources rather than focusing solely on one of them. This means, for example, that language or discourse, while considered an important element, is not given the primary or sole focus, unlike in many sociolinguistic and discourse analytical works, including some of my own. There are, of course, other concepts and frameworks that emphasize the importance of the inclusion of various dimensions. Recent discussions on posthumanism (Braidotti 2019; Hovens 2023; Lamb and Higgins 2020; Pennycook 2016) and on the Anthropocene (Chakma et al. 2023; Chakrabarty 2018; Cornips and van den Hengel 2021; Haraway 2015; Lamb 2024; Tsing et al. 2017) emphasize the importance of nonhuman forces and actors in the conceptualization and examination of current complex processes and changing conditions, such as the value of natural resources, climate change, and social justice.

The material turn in humanistic and social sciences has embraced the concept of assemblage because it explicitly includes materiality while criticising discursive approaches of disregarding the material aspects of social, political, and economic processes (Bennett 2010). However, both Foucault, and Deleuze and Guattari insist that both aspects—material and discursive—are fundamental to understanding the social world and that they are often intertwined, thus creating together the force of a particular constellation or discourse. Another important critique regarding the work by Foucault, as well as Deleuze and Guattari, comes from postcolonial thinkers, such as Said (1995) and Spivak (1988) who point out, for example, the lack of critical reflection on ideology and epistemic positionings of their work, as well as a too limited space for self-reflective activities of subjects as well as a lack of explicit engagement with postcolonial issues and thought (for details, see, e.g., Bignall and Patton 2010; Nichols 2010). At the same time, the work by Foucault, as well as Deleuze and Guattari has also been applied in postcolonial studies related to, e.g., ethics of the self, multiplicity subjectivity and critique of capitalist and colonial processes (e.g., Buchanan 2021; Mahmood 2012) and some research has documented their interest and active participation in various political movements, including struggles for decolonization (Dosse 2016; Bignall and Patton 2010).

Assemblage also needs desire to function. Deleuze and Guattari (1980/1987) argue that desire is the engine that powers assemblage, giving it its affective purpose and anchoring it to people. This resonates with Foucault's writings (1982) about self, bodies, and biopower, which highlight the impact of power techniques both on the conditions of knowing and on the conditions for situated subjectivity, allowing individuals to navigate a messy terrain of constraints and possibilities. In assemblage thinking, the affective dimension underlines the importance of experience and desire when thinking about revaluing Arctic natural resources. Resonating with the Spinosian understanding of the concept, affect places emphasis on embodied experience (Ruddick 2010). In this book,

I apply the Deleuzian understanding of affect to refer to an embodied experience that is always grounded in a specific situation and bodies. This makes ethnography a productive method for examining affective aspects of the Cold Rush assemblage (see Chapter 2).

These concepts aim to capture the multiplicity and process-like nature of the world. They resonate with the critical argument by Foucault, and Deleuze and Guattari, who reject the idea of one overarching, general truth about or essence of anything. Rather, they argue for a multiplicity of every "thing": there are always alternative ways to assemble and arrange elements for a particular assemblage and multiple discourses related to a particular event or thing. Translated to the Cold Rush terrain, this means that there is no single "real" or "authentic" value or function for natural resources, but rather a variety of value constructions, some overlapping and many intertwined, some more enduring, others more quickly changing. Deleuze and Guattari's concepts of assemblage and rhizome, as well as Foucault's notions of discourse regimes and order of discourse, attempt to capture the politics of these connections, relations, and multitudes and how they work together under particular conditions of a given time. All these concepts place power as a central force in how assemblages, rhizomes, and discourses function. They are all deeply critical and political concepts: they impact the object of action, the participants, and the relations between them; they are not organized by accident or by laws of nature but by particular interests and power relations. The particular organising elements in an assemblage are a matter of power relationships: nothing is inherent in how people, natural resources, or the environment are categorized. It is a power game. Assemblage, rhizome, and discourse are part of a bigger picture; they can be used to stabilize or destabilize dominant ways of thinking and acting.

The rejection of a singularity emphasizes the need to account for temporal and spatial variations of revaluing natural resources. A wealth of anthropological research has shown how natural resources are revalued by turning them into products, properties, and social relations (Ferry

2013; Narotzky and Besnier 2014; Tsing 2015). As a social construct, value is made and re-made in interaction under particular conditions and in a particular time and place. Seeing value as a dynamic process highlights how the value of natural resources can vary across time and place or how resources can have different values simultaneously. Viewing value as a constellation of actions that produces things and qualities, not as a static quality in itself, we can see how all sorts of motivations, alliances, associations, and interests intervene to create value (Ferry 2020). This also resonates with assemblage thinking in examining the revaluing of Arctic natural resources. The revaluing process can be productively thought of as an assemblage of material, discursive, and affective elements that together produce a situated understanding of the value of the natural resources embedded in ecological, political, economic, and cultural relations and processes (Bazzul and Tolbert 2017; Pietikäinen 2021).

In addition, value dynamics highlight the tension between material objects and the social relations they express (Narotzky and Besnier 2014; Shankar and Cavanaugh 2012). Value-making practices are always embedded not only in a specific time and place but also in social relations that not only produce value but also create people as locals and tourists, for example. Revaluing natural resources has a long history of colonial and postcolonial extractions, future imaginations, and speculations. What makes something valuable is its capacity to preserve, increase, or transform its worth as it moves through time and space (Peebles 2010). Consequently, we can follow objects through multiple regimes of value and, in doing so, seek to show how value is made and remade (Ferry 2013). This movement can be examined as the dynamics of de- and reterritorialising processes, which, as Guattari and Deleuze (1980/1987) remind us, are a core process in capitalism.

1.3 Arctic Cartographies: Mapping the Political Economy of Land

Fig. 1.3 Arctic circle map (*Source* Hugo Ahlenius, UNEP/GRID-Arendal)

The cartographies of Finnish Arctic extraction look different depending on whose map we are looking at (Fig. 1.3). For indigenous Sámi people, the northernmost tip of Scandinavia, including part of the Finnish Arctic region, belongs to their traditional homeland, *Sápmi* in the Northern Sámi language. The land was first used for gathering, hunting, and fishing, and later, especially since the seventeenth century, for reindeer herding and, gradually, for other livelihoods, including farming and tourism (Lehtola 2012). The area has always been a place of mobility and contact related to the seasonal movement due to livelihoods as well as trade and indigenous governance. The indigenous *Siida* system (a network of winter villages) served as the foundation for communal life and governance of natural resources. People living in the villages had exclusive rights to use land and water areas within the boundaries of their respective *Siida.* At the beginning of the Modern era, thirteen *Siida* villages existed in the territory of present-day Finland. The Swedish Crown and Russian/Novgorod rulers both used village systems to govern the territory and sent their bailiffs to these villages to collect taxes. Still today, many of these traditional *Siida* villages are now central service centres for the region. The *Siida* system remains relevant to the region's current debates over landowning rights (Lehtola 2002; Tervaniemi and Magga 2018). While there is an ongoing legal debate about the extent to which the Sámi owned the land they were using in the modern sense of ownership, many scholars (Joona and Joona 2023; Kuokkanen 2023; Lehtola 2012; Valkonen et al. 2017) argue that the governance of the land, organized around the *Siida* villages, had already established ownership.

Since the seventeenth century, expanding settler colonialism has transformed Arctic governance and relationships between people. As a result of colonial practices, the Sámi people were constructed as nomads without a legal claim to land and water. Instead, the legal ownership was distributed to the Crown and the settlers (Kuokkanen 2022, 2023; Lehtola 2021). This restructuring of relationships around the Arctic land and waters continues to impact the current political, legal, and economic relationships around them. Today, the Sámi people in Finland are legally recognized as indigenous people, but this recognition only grants them cultural rights, while the land and water rights are still not settled (Kuokkanen 2023; Valkonen and Valkonen 2017; Valkonen 2018).

Historically, the Finnish Arctic was considered an uncharted territory at the end of the world. Prior to the sixteenth century, written historical accounts of this region and its people are sparse and typically made by visiting explorers sent by southern centres. In the few surviving written accounts, the region is typically depicted as the ultimate periphery—the far end of the world with no trace of culture or civilization. In classical European literature and maps, the region is called *Ultima Thule*, often considered an island in antiquity and, in modern interpretation, assumed to refer to Scandinavia. The oldest written description of the Arctic region and its people is by Rome Cornelius Tacitus in Germania (in year 98). At the time, southern rulers did not have much interest in the region or its people. The borders of the kingdoms of Sweden and Russia were not marked in the territory but rather politically agreed upon on paper between the Swedish Crown and the Russian Tsar for taxation purposes.

Until the beginning of the Modern age, the southern rulers described the region in official written documents as no man's land *terra nullius*, hostile, empty, and cold. This historical construction of the Arctic as a wild, unruly place recirculates in the construction of the Arctic as a frontier, available to be grabbed for economic exploitation. Explorers sent by kings, rulers, and churches amplified these ideas, especially since the sixteenth and seventeenth centuries (Spangen et al. 2015). Similar to other areas constructed as frontiers (Bebbington et al. 2014), these explorations spatially charted the territory deemed free for extraction and ignored or erased existing land-use systems.

Since the seventeenth century, Southern actors' growing economic and political interests have focused on this Arctic region. At this time, the settler colonialism intensified (Kuokkanen 2022, 2023; Lehtola 1997). One of the persistent motivations is having access to and control over natural resources, especially water and land. For this reason, the Swedish Crown began to increase its control over the region's spatial, economic, and political re-organization. Lehtola (1997: 39) argues that this interest materialized in three interrelated ways: through religious conversion, state governance, and a settlement policy, all forms of colonialization and capitalism at the time. The interest in controlling and capitalising Arctic resources was linked to the wider shifts in political and economic rationalities at the time. The overarching rationale is modernity, understood here, following Foucault (1972/1977), as a historical category marked by developments such as faith in progress and rationalization, a shift from agrarianism towards capitalism, and the rise of the nation-state. Foucault's critique of modernity is centred on power/knowledge regimes that govern historical periods and what is understood as "truth" in terms of the logical, best thing to do. These rationales sometimes worked on their own but were typically interlinked, overlapping, and minimally echoing one another. Following the temporal order, they became relevant for the governance of the Arctic, and by the driving forces fuelling them, these rationales can be called progress, nationalism, and globalization.

After the late seventeenth century, the modernization project started to take a firmer hold in Lapland, and it was carried out in the name of progress. It brought with it several overlapping and interconnected changes, driven by the new economic interest of the Swedish Crown, which needed to expand its economic base and increase tax revenues. To ensure the resources for economic growth, the Crown strengthened its administrative and ecclesiastical ties in the North. The strategy was to establish new regional governance, allowing reconfigurations of economic relationships.

A powerful tool for this new governance was the creation of a new category of settler (*uustilallinen*) and the settler farm (*uudistila*) not only for taxation purposes but also for reorganising land use and related rights. In this context, settler/settler farmer refers to a person with a house and some surrounding land for small-scale farming. The category settler/settler farm introduced a politically consequential and economically invested shift from ethnic categories (Sámi, Finns) to an economic category of livelihoods in relation to both the state/crown and the land. It imposed the temporal erasure of historical developments in the governance of the land and people by constructing and imposing the ahistorical, economically invested category of the settler. This created a clean slate for early capitalism in the Arctic. The settler category was granted by the Crown; it was open to both Sámi and Finns, erasing the temporal difference between the newcomers and those who had long resided in the area (Kuokkanen 2022, 2023; Lehtola 2012, 2021). At the time, Sámi made up the majority of the population (Lehtola 1997). Under the changing conditions, it made economic sense for many Sámi to subscribe to the category of settler farmer and register their house and lands as a settler farm, regardless of how long the family had lived and used the land (Spangen et al. 2015). 90% of all new settler farms established in the seventeenth century belonged to a Sámi family (Lehtola 1997: 43).

This can clearly be seen as a process of colonialization (Spangen et al. 2015). In becoming a settler within this system, Spangen et al. (2015) argue that the Sámi became “finlandialized” not only in terms of livelihoods but also culturally, including linguistically. While the settler category was open to anyone, albeit rooted in the majority system, subscribing to it meant becoming recognized as a Finn and backgrounding, at times even erasing, the Sámi roots. Settlers began to sound like Finns; Finnish became the dominant language, with potential for social mobility in the state-driven Arctic. Many families were given or adopted a Finnish family name, typically the geographical characteristic or the name of the place they lived. This economic restructuring also entailed a cultural transition, echoing similar kinds of colonial processes in other Nordic countries (see, e.g., Andersson 2020; Graugaard 2019; Maegaard and Mortensen 2022; Outakoski 2022; Sollid and Olsen 2019).

Two interconnected discourses were adopted and circulated in the public domains to legitimize the economically driven redistribution of resources. The first introduced the idea of “parallel theory”, as a way of explaining the logic and benefits of the new governance of the land and people in the Arctic. According to this theory, the two ways of using land, the traditional indigenous Sámi way and the modern Finnish way, were distinctive and different. Thus, they were not in competition with one another, nor did they disturb one another; rather, the vast Arctic region accommodated both parallel uses in parallel harmony. The second discourse, describing the Arctic as a frontier, keeps the historical idea of the Arctic as a no man’s land in circulation, ideologically legitimising the colonialization of the Sámiland and exploitation of the commons. Both discourses are still circulating today. For example, the parallel theory discourse is evoked when discussing Lapland’s competing and conflicting economic interests in the tourism and mining industries (Jokinen 2019).

Similarly, national cartography locates the Arctic region on the periphery of the nation-state, depicted as a reservoir of people, resources, and space for the centre's economic development (Pietikäinen and Kelly-Holmes 2013). In the early twentieth century, following the national logic of a newly independent Finland, vast Arctic areas were claimed and categorized as state forests, primarily for the timber industry, earning the nickname "green gold" in post-World War II Finland. Even today, the State of Finland still owns significant part of the forests in Lapland. According to Finnish law, the same forests are commons, and everyone has been granted the so-called every person's rights, which include free access to forests, lakes, and rivers and the right to pick berries, mushrooms, and unprotected plants, provided that this does not harm nature or interfere with the landowner.

Indigenous groups across the Arctic have contested these colonialist practices of deterritorialization and erasure, countering that the region was never an empty terrain but rather a thriving land of indigenous livelihood and governance (Junka-Aikio 2023; Kuokkanen 2019; Valkonen et al. 2022). To overcome erasure resulting from settler colonialism and nation-state regimes, as well as to contest the continuing process of turning indigenous homelands into frontier zones for extraction (Bebbington et al. 2014), several Arctic indigenous peoples have engaged in recognition and reconciliation processes, with varying degrees of success (Coulthard 2014; Kuokkanen 2022; Simpson 2017). Political recognition, however, does not necessarily entail resource redistribution as it can be limited to cultural governance and self-determination (Povinelli 2011). This is the case for the indigenous Sámi people and their homeland, *Sápmi*, which stretches across the northernmost tip of the Arctic in Norway, Sweden, Finland, and a corner of Russia's northeast region. For example, in Finland, the Sámi have cultural self-governance but no special rights to land or natural resources, and the political struggle to change this is complex and ongoing (Joona and Joona 2023, Rantanen and Valkonen 2011; Valkonen 2018).

The Finnish Arctic is a growing economic hub: in 2021, statistics show that over 3.1 million tourists stayed at least one night in Lapland, which is home to about 200,000 people (www.lapland.fi/business, accessed 11.8.2023). Tourism brought nearly 767 million euros in 2021 to the region, which is losing jobs in primary sectors, especially in farming and timber (Lapinluotsi 2022). The mining industry has long historical roots in the region, and the combination of new technology, renewed infrastructure, and the impact of climate change on permanent frost has made mining investments in Lapland lucrative. For example, in 2021, the total turnover of Lapland's mining industry was 912 million (Lapinluotsi 2022).

A more recent economic interest focuses on the economic potential of Arctic berries, mushrooms, and herbs for global markets chasing the superfood trend, supported by growing interest in local, ecological, organic, and traditional foods, ingredients, and recipes. Underlying much of this economic development is another kind of economic system known as project economy. It is extensively supported by a subsidy system provided by the municipalities, the Finnish government, and the EU, especially in areas, such as Lapland, that are classified as peripheral. With expanding economic activity, the concerns for ecological and cultural sustainability and well-being are growing. The challenge of extracting and protecting Arctic resources extends beyond nation-state borders, transgressing territorial, cultural, and linguistic boundaries and triggering conflicts over legitimacy, governance, and profit distribution.

These different Arctic cartographies create "overlapping geographies" (Bebbington et al. 2014: 22), each one attempting to define, lay claim to, and regulate particular places as resource spaces as well as determining who should have access, use, occupy, govern, and control them. Various maps chart, for example, the prospective mining area on top of protected areas, which in turn have been mapped on indigenous Sámiland. More than just spatial overlaps in a simple descriptive sense, these overlapping geographies can be seen as contradictory political and economic projects (Keskitalo 2019). They become part of the power struggle over the control and use of Arctic natural resources and different ways of occupying and governing resource spaces.

The Arctic has been territorialized for specific purposes, including mining extraction, natural parks, state forests, tourist resorts, service centres, and indigenous land (Nuttall et al. 2018). It also functions as a site for reproducing and changing the dynamics of colonial and nation-state logics, indigenous politics, and globalized capitalism (Kuokkanen 2019; Schweitzer 2022a, b). These diverse, often incompatible, interests and values have turned the Arctic into a contact zone (Pratt 1991) or even inevitably a conflict zone between national interests, global economic investment, ecological concerns, and the struggles of local communities, including several indigenous nations across the region.

The Finnish Arctic is a laboratory for these multiple, ongoing, and overlapping processes. Depending on time, place, and the narrator, the region is referred to as *Ultima Thule, Lapponia, Lapland, Finnmark, Sámiland*, and *Sápmi* for example. Depending on a particular era's political and economic interests and its stakeholders, people living in the area have been categorized as *Fenni, Lapps, Sápmelas, Sámi, Finns,* and *Laplanders*. Recent political and economic interest in the governance of Arctic natural resources has stirred up new interests and old tensions around identity categories that might grant legitimate and better access to these resources (Junka-Aikio 2023; Valkonen 2009; Valkonen and Valkonen 2017; Sarivaara 2012).

In this book, the Arctic has multiple meanings. It refers to a rhizomatic and emergent space that simultaneously encompasses (1) the indigenous Sámi homeland, *Sápmi*, geographically located in the arctic Scandinavia, historically existing and politically (re)constructed under the cultural self-governance of legally recognized indigenous people; (2) Lapland, a northern periphery under the logic of the nation-state, now revalued as an emerging economic hub under globalization; and (3) the Arctic region as a flexible and shifting construct with special associations for the Nordic countries, circumpolar communities and even the Northern hemisphere, often referring to the geographical features and natural qualities of the area.

The Arctic thus refers to a shifting, ambivalent, emergent space characterized by intertwined processes of identity politics, changing state restructuring, emerging economic development, and accelerating climate change. This conceptualization distances itself from a geographic notion of space as a fixed, unchanging vacuum in which we locate individuals, objects, or events; instead, it frames the Arctic as a discursively produced space with material conditions and consequences—some of which may indeed be geographical—and with temporal as well as spatial aspects. In this view, space is a dynamic social construction, its boundaries subject to constant negotiation and redefinition. What we call the Arctic is carved out and performed in a complex and changing system of material practices and circulating discourses, each with its own history and affordances for particular actions. In Foucault's term, this view emphasizes that we live in heteroglossic spaces, or heterotopia, where structures and counter-structures, logics, discourses, and regimes co-exist, interact, overlap, and struggle (Foucault 1977/1984). This view of the Arctic reflects a complex sense of place, mobility, and community, and it tells a changing story of fixed and fluid ideas of natural resources, legitimized access, and ownership.

Today, the Arctic is a high-stakes geopolitical region with diverse histories and communities, accelerating climate change, and speculative future imaginations related to extractive industries, shifts in global politics, and sustainability challenges. This multiplicity creates a dynamic space marked by competing ecological, economic, and political processes. This dynamic complexity is influenced not only by ecological change, the political movements of indigenous nations, historical developments, settler colonialism, nation-state regimes, and supranational policies but also by flexible and mobile forms of capitalism. These developments have increased the region's strategic importance while prompting new ideological confrontations and forms of economic investment. Furthermore, economic processes develop unevenly and differently in historically specific places, and they are neither totalizing nor seamless. Neither are these processes simple or linear. Instead, they must be understood as rhizomatic, incomplete, and ongoing. In this book, I try to capture and analyse this multiplicity with the help of critical assemblage analysis (see Fig. 1.5).

1.4 Arctic Boom, Bust, and Buzz: The Temporality of Natural Resource Extraction

Fig. 1.4 The forest taking over the old railway track to the mine

The spatial reconfigurations of the Arctic are intertwined with its extraction temporalities and the enduring Arctic rhythm of each of the four seasons. Each differs in weather conditions, but together, they also set the rhythm for activities and work. For example, the Arctic berry temporalities are affected by weather conditions, which impact how they survive the cold winter, pollinate, and grow and ripen over the very short Arctic summer. Due to annual variations in growing conditions, the berries do not show up in the same locations every summer, so finding the best berry patches for each season requires a combination of traditional knowledge, long walks in the forest, and a little luck. This cycle impacts the value of the berries: a good berry summer means prices are low, and the reverse is also true. In summers, when berries are hard to find, their value increases. Nature-based industries, including farming, forestry, foraging, and tourism, typically have one or two peak seasons when much of the workforce is needed. Some industries, like mining, operate all-year-round (Fig. 1.4).

Another prevalent Arctic temporality is the cycle of boom, bust, and buzz of economic development, which is mainly related to the life cycles of tourism and mining industries.

The temporal dynamics of extractive industries include a relatively short-lived active mining period, followed by a bust: the closing of the mine and the emptying of the mining town, which has ripple effects (Bell 2023; D'Angelo and Pijpers 2018; Halvaksz, 2008; Pietikäinen and Allan 2021). Living near the ruins with echoes of the boom fuels occasional attempts to re-open the mine. The dynamics of mine opening and closing are an ongoing feature related to fluctuations in the price of minerals (Pietikäinen 2019). Like many other geopolitical peripheries, the Arctic is vulnerable to global changes due to a lack of an extensive, diversified economy or reservoir of surplus labour (Pietikäinen et al. 2016).

Past, present, and future temporalities invest the Arctic with nostalgia for historical developments, the urgency of climate change, and the rush for resources (Ahlness 2019). The beat of speculation, the most pervasive economic temporality in the Arctic, patterns these temporalities. Bryant and Knight (2019) argue that central temporal orientations towards the future include waiting, hoping, anticipating, and speculating. Among these temporalities, speculation makes the future malleable and manageable by calculating odds, opportunities, and risks. Speculation entails risk, but hazarding the right choices lays the groundwork for future returns. Speculation is an impetus for the Cold Rush: "Now" is the time for the Arctic, with its vast resources becoming more accessible as climate change progresses. I use the metaphor of fever, e.g., gold fever, fishing fever, and berry fever, to refer to a moment when the commodification of commons such as these is fuelled by a sense of urgency that compels people to drop other tasks and rush into action before others arrive. Fever is also short-lived and intense; it burns itself out.

Speculation is an enduring temporality of capitalism. Bear (2020: 1) describes speculation as future-oriented affective, physical, and intellectual labour intended to accumulate capital for various ends. It proceeds by making value uncertain and then projects various strategies to help navigate the uncertainty. Speculation technologies seek to turn ambivalent data, dispersing signals and pools of information into better knowledge, impacting the course of action (Hong 2020). These technologies work through discourse, multimodal explanations, future-oriented narratives, and affective language. They also shape current actions, such as granting drilling permissions or mineral expeditions in the natural reservoir areas. They affect branding strategies, political roadmaps, value calculations, and risk analysis, and they are used to govern and anticipate the future (Bear 2020). Dematerialized speculative bubbles cause a crisis (Kerslake 2015), which initiates yet another cycle of speculation about the nature, effect, and length of the crisis.

Like weather broadcasts, speculation becomes intertwined with everyday Arctic business: when will the tourists come, berries ripen, snow fall, and Northern Lights appear. Ferry (2020) develops an idea of speculative substance and argues that, for example, in oil exploration, speculation becomes significant through exploration technologies such as drilling. The same applies to Arctic mining test drillings. Depending on the direction of the speculation, the prognoses can be boom or bust, crisis or profit. Speculation works rhizomatically; the transnational links to the Arctic natural resources assert that we must always think of "here" in relation to a "there" (Bear 2020: 6). In the Cold Rush context, speculation can be used to reason or challenge economic projects, current governance of commons, and infrastructure investments.

At the heart of economic speculation is a trust in ongoing growth. This trust was abruptly shaken in 2020 due to the global pandemic. At the time, more than half of Finnish tourism companies operating in the Arctic region were forecast to go bankrupt in the next season as markets rapidly shrank. All these risks did not materialize, and as the pandemic finished, the new boom cycle began again.

1.5 Writing Cold Rush: Politics and Ethics of Assembling Meanings

Fig. 1.5 The salmon and the fisher

I have become a salmon widow every summer for the last thirty years. This is a local term for a non-fishing partner in a relationship, where the other partner is "lost" to a short-lived salmon fever: intensive, frantic, all-consuming fishing activities, excluding everything else except fishing, or talking about it or preparing to go fishing or recovering from it. The salmon fever usually lasts from a few weeks up to a month, depending on how long we stay with the family in the Tornio riverside village where my husband, a professional photographer in our southern life, is from. The fishing is all-encompassing. People, activities, and places are organized around binary evaluative categories valuing fishing over everything else: fishers-others; fishing-other activities, and good fishing site-other sites. The rhythm of the days is organized around fishing, especially around the speculative best moments when the fish is supposed to bait, demanding immediate fishing activities. To me, it seems to be always. During the first summers, I looked at this frenzy of activities half amused, mostly in disbelief and with growing impatience. It became clear to me that I was new to this salmon fever and an outsider to the region. In the beginning, my mother-in-law tried to train me to become a good salmon widow, which seemed to include a lot of waiting for the fishers to return, keeping the food and sauna warm for them, and patiently listening to their endless, repetitive stories of how they almost got the king of the fish. Gradually I became more tolerant towards this imposed salmon widowhood and started to see it as an opportunity to do my own activities. To my surprise, I became drawn into watching as the salmon fever ran through the extended family and the whole village where we lived in the summers. Eventually, I realized that all this was more than just fishing: it was about being part of the community, creating links across the network of neighbours, a centuries-long tradition, and the Tornio river. It was a mode of belonging (Fig. 1.5).

As I gradually moved from being dismayed about the fishing dominating our Arctic summer practices to listening and watching how the fishing and stories about it unfolded over the summers, I noticed patterns, entanglements, and paradoxes. Fishing is not only connected to cultural local practices but also to the well-being of the whole local river ecosystem, which is connected to the well-being of the ecosystem in the Baltic Sea and in the Atlantic Sea, i.e. the waterways the salmon travels through. Importantly, a fish's well-being is connected to the well-being of the rivers and seas that create its waterways. The fish and the river become intertwined with the mining industry's prospects, as it uses the rivers in its production processes. Mining has over 400 years of history by the Tornio River, and many locals have a story to tell about the former glory days of the mining industry in the area. Many also worked in some of the local mines in the mining boom in the 1970s and 1980s, and a few works in the mines that are still operating in neighbouring communities. The mining industry's past, present, and speculative prospects connect the river to the global colonial and capitalist processes, including past expeditions sent by southern kings or current expeditions by multinational mining companies.

The Tornio river connects the villages on both sides of the river bank to the local identity of *Torniojokilaaksolainen* (one who lives in the valley of the Tornio River), while, to outsiders, the river is an official border between Sweden and Finland. Spending longer periods of time every year in this particular borderland region, where we now have our second home, I have gradually learned about the histories of economic development related to mining, timber, farming, and tourism. The changing but continuing relationship between the Swedes and the Finns, the Sámi and the Finns, locals and outsiders, northerners and southerners, river villagers and lake villagers, and newcomers and locals are all among the categories evoked and narrated in stories of economic, political and cultural changes, tensions and continuities. During the past three decades that I have lived in the village part of the year, I have seen several contradictory yet simultaneous processes taking place in the region. The old village school building has been transformed into fishing tourism destination. The closed mines, initially owned by the State, have been bought by international investors with a promise to re-open them. The value of mines has become more ambivalent, including the traditional understanding of mines as a source not only of progress and permanent jobs, but also of pollution, eco-crises, and harm to tourism and fishing. The arrival of imported Thai berry pickers in the region has now become a sign of the start of the berry harvest season. During this time, some parts of this Arctic periphery in the Finnish state system turn into an economic hub for global tourism, mining, and wild food, while other parts are losing their final welfare services and inhabitants. My husband's childhood village has fewer and fewer permanent inhabitants. Nowadays, when staying at our house in the village and I walk my dog along the snow-covered village road in the dark winter evenings, I can see only a few lights shining from the windows of a few houses. The rest are hibernating until the salmon and the former villagers return next summer. The periphery, too, has its centres and margins, and the dynamics of utopia and dystopia and their temporalities are palpable. The Cold Rush is an ambivalent phenomenon, simultaneously holding hope and harm for the Arctic environment and its people.

These connections, entanglements, and contradictions sparked my interest in developing a research project examining the Cold Rush. I wanted to understand better the possibilities and tensions related to the revaluing processes of natural resources and how people cope and strategize with these changes and possibilities. The acceleration of the rush materialized in the tourism and mining boom, which, together with the acceleration of climate change in the Arctic regions, stirred existing practices and logics of action and challenged entrepreneurs, municipalities, and locals to reevaluate their current and future strategies. An ethnographic approach meant that I observed people in their everyday practices and listened to the stories that people were telling about these changes and possibilities, be they past, present, or speculative, and reflected on the situation together with the participants. A discourse analytical approach helped me to look at how people were making sense of the situation and their actions and to map the trajectories of the rationalities evoked in their stories. Both discourse analytical and sociolinguistic approaches oriented me to look closely at the relationships between meanings and broader social, political, economic, and ecological processes and structures. The experience from multi-sited ethnography, analysis of circulating discourses embedded in power relations, and mapping them onto the dynamics of language and society led me to look for concepts that would help me to grasp and make sense of the multiple, partly overlapping connections, interactions, and entanglements. I found rhizome and assemblage.

I made a conscious attempt to design the Cold Rush project with the concept of rhizome in mind. I have previously worked with nexus analysis (Scollon and Scollon 2004) and multi-sited ethnography (Pietikäinen and Kelly-Holmes 2013; Pietikäinen et al. 2016; also Heller et al. 2018). However, I now wanted to imagine a project that would emphasize the connections and interdependences among multiple and complex processes. As a result, the Cold Rush project design had multiple entry points, some related to expanding extractive industries (mining, tourism), some related to bodies (mobile workers, locals, miners, family entrepreneurs), and yet others to land (berries, minerals) and infrastructure (airport, snow, air, rivers). Some of these entry points were anticipated based on the previous research, my pilot studies, and long-term observations of regional changes, but also new Cold Rush entry points emerged as the project developed. The rhizomatic design allowed multiple, previously unknown, entry points to the Cold Rush processes and the field and sites where natural resources were being revalued. Furthermore, as a multi-sited and multi-temporal ethnography, the project was also an example of "always-in-process collaboration" (Tsing 2015: ix) between the team members working for different lengths of time on particular aspects of the Cold Rush. Team ethnography was an important part of the project, especially at the beginning when we were mapping the potential entry points and key economic processes. In practice, we visited numerous (possible) sites with varying combinations of team members and collaborators, and wrote shared fieldnotes based on our observations and conversations relevant for the economic transformation of the Arctic. Some of the entry points developed into particular case studies carried out by the team members and collaborators (see, e.g., Strömmer 2021; Allan 2019; Ojala 2020), while other team members, especially Monica Heller and Alexandre Duchêne, contributed to the theoretical and empirical development of the project.

This book, however, is based on two entry points that I have to the Arctic and the Cold Rush: to my ethnographic fieldwork on the Cold Rush in the Finnish Arctic between 2016 and 2022 related to mining, Arctic wild berries, and the Northern Lights business, on the one hand, and to my over 30 years of experience of living in this region part of the year, on the other. The emergent, intertwined nature of the rhizome resonates with the emergent nature of fieldwork: I mapped and followed the Cold Rush transformations, or attempts or speculations of such transformations, related to land and water (fishing, mining, wild berries) and air (Northern Lights). I did most of this mapping alone, but a few times, some team members accompanied me to an Aurora hunt or mining sites. Over the six years, my fieldwork evolved and created a rhizome where many lines became connected, a few were interrupted, some were dead ends, and some sprouted out to form new lines and connections. This rhizome forms the empirical basis of this book (see Chapters 3, 4, and 5).

In my research, I focused on historical, present, and speculative developments of mining and berry industries in the region as well as novel convergences of commons (land, water, air) into commodities in tourism as well as in various economic development projects related to these natural resources led by the municipalities. In practice, I did archival work (e.g., historical records, media archives), interviewed stakeholders (municipal decision-makers, workers, tourists, entrepreneurs, and local people), and observed meetings related to the economic development of the region as well as people experiencing the Cold Rush by living near to tourism resorts, mines, or the Tornio river. While doing this fieldwork, some people I met and interviewed became key participants in the research as they were engaged with the transformations I examined and willing to share their experiences. You will learn more about these key participants in the empirical chapters of this book.

The book also resonates with my previous work in the Finnish Arctic since the late 1990s. This has entailed several research projects, starting with my Ph.D. dissertation (2000) on the discursive construction of ethnic difference in Finnish news discourse and a post-doctoral project on indigenous Sámi language media in the 2000s. Before the Cold Rush project, I led two multi-sited, critical ethnographic, sociolinguistic, and discourse analytical research projects in the area focusing on identity and multilingualism (Northern Multilingualism 2008–2011, see Pietikäinen and Kelly-Holmes 2013) and multilingualism and economic and political processes (Peripheral multilingualism 2011–2015, see Pietikäinen et al. 2016).

The story of the unfolding Cold Rush is both hard and easy to tell. The multiplicity of the Cold Rush processes, a range of relevant concepts, and rich ethnographic data translate to multiple choices as I attempt to craft this multiplicity into a text that will make sense to you. I believe that writing, too, can be thought of as an assemblage: it is about selection and bringing into interaction previous theories, scholars, data, and analysis with a particular way of writing and organizing the parts of the book to produce a productive interaction between research, writing, and reading. Writing assemblage also entails the politics of writing, linked to the argument being developed, the rationale for choices made, and the principles of organization. In a book like this that encompasses multiple concepts from various disciplines as well as involving a region that is the object of multidisciplinary research, the potentially relevant previous literature is vast. From this vast literature, I have selected those that have helped me to think further in-depth about the processes of revaluing of the Arctic Natural resources in the Finnish context. There are excellent books that provide introductions to the thinking of Foucault, and that of Deleuze and Guattari (e.g., May 2005; Buchanan 2021) as well as ethnographic research on Arctic regions (e.g., Demuth 2020) and history and politics of the indigenous Sámi people in Finland (e.g., Kuokkanen 2019; Lehtola 2022; Valkonen et al. 2022). At the same time, my reading choices are limited by my linguistic skills: I have been able to read publications in Finnish, English and Swedish. By the time I finished this book, new AI translation tools were emerging, which can potentially become useful tools in extending the reading for future projects.

I have chosen four entry points to examine the Cold Rush: one theoretical (Chapter 2) and three empirical (Chapters 3, 4, and 5). In the theoretical chapter, I discuss in more detail the key concepts of this book—assemblage, rhizome, and discourse—and how they together form the basis for a framework to examine the unfolding of complex processes and the interaction between their material, discursive, and affective aspects. I also discuss the connections between this framework and ethnography. The three empirical entry points each lead to a hotspot of the Cold Rush: mining, Arctic wild berries, and Northern Lights. These are the hotspots where I have done most of my longitudinal fieldwork over six years, and together, they represent an interconnected rhizome of the revaluing of Arctic natural resources. In Chapter 3, I discuss the historical trajectory of mining in the area and how it became intertwined with colonialism and capitalism and came to be seen as progress. In Chapter 4, I focus on revaluing mundane berries as gifts, work, and superberries. The berries remain biologically the same, but they gain new meanings and values through discursive revaluing. Chapter 5 shows the unfolding of the revaluing of Auroras: how driving around in a pitch-dark Arctic night with a skilful guide who has great stories becomes a unique nature experience for tourists and a successful business for entrepreneurs. The concluding chapter (Chapter 6) discusses what happens if we look at these kinds of processes, like the Cold Rush, with the help of the concepts of machine and production rather than structure and representation.

Following the rhizomatic design of the Cold Rush project and resonating humbly with the example of the rhizomatic book structure of the Thousand Plateaus, one of the main contributions by Deleuze and Guattari, this book can be read in a flexible order. While the introduction and discussion probably make sense to read first and last, any other chapters can be an entry point to the Cold Rush, which can be read independently and in whatever order you wish. I recognize that this kind of structure entails a risk of becoming a confusing entity lacking the clarity needed to engage with it. At the same time, this approach opens up a space of creative and critical thinking and a stance for me as a researcher and as a writer. It has taken me a long time to write this book, and it may well be that reading this book takes time: to think, to linger on, and to create your own connections.

While I do value clarity in thinking and writing, I also recognize that clarity is one type of writing ideology materialising, for example, in norms such as preferring linear (tree) structure, unambiguous definition, and lack of repetition. I would like to think that there is also a rhizomatic alternative to reach connections and understanding between the writer, text, and reader, and this is the way I set out to do in this book. This rhizomatic stance also resonates with the languages I have used when working on the Cold Rush project. Most of the interactions and the written texts were in Finnish, some in English, and occasionally in Swedish, Meänkieli, or Northern Sámi. I write this book in English, which is simultaneously an easy and complex decision to make. It is easy in the sense that English bridges with readers beyond the Finnish-speaking community but difficult in transforming interactions and local knowledge from one relatively small language to another global lingua franca. English is a working language for me, which makes it easy and difficult to write with. It is difficult as I need to navigate the terrain of norms, standards, correctness, and accessibility, together with my desire to write with a voice that I recognize as my own even if it is not in Finnish. At the same time, writing in English also feels freeing as it does not hold a burden of "native-ness" for me. However, the rhizomatic structure together with my shifting, in-between position across categories related to languages, places and identities creates a "smooth" space in Deleuze-Guattarian terms, a kind of a twilight position that allows me to think and write with smooth eyes and hand.

The ethnographic Cold Rush photographs accompany the text in the book. I used photography as a way to document how the Cold Rush unfolded over time across seasons and empirical hotspots, yielding to nearly 5000 photos of the Cold Rush. Photographing is a way for me to make field notes, remember, and reflect. Photographing was also a key element in the *Cold Rush Science* postcards, designed and distributed to convey some of the project's key findings. Resembling art postcards featuring selected artwork at the project's exhibition, the science postcards provided a way to discover and articulate the entanglements of the Cold Rush findings, processes, and hotspots in a genre that typically indexes place, mobility, and greeting. These snapshots were designed to tease affective responses from the receivers and to invite our stakeholders, students, and colleagues to consider the Cold Rush and its ramifications. These postcards were an alternative to a final written report and some are used in this book.

The production of the science postcards was designed together with Alexandre Duchêne, and it took over two years and entailed several steps. Together with the professional photographer (that salmon-fishing husband of mine!), I visited the key Cold Rush sites in all four seasons for over two years (2019–2022), and we took pictures of the same places, events, and processes, which yielded over 3000 photos. This period of time also included the global pandemic, which became intertwined with the boom, bust, and buzz aspects of the Cold Rush unfolding in the region. With the help of a research assistant, Hilla Rötkönen, our oldest daughter who recognized the places, the photos were categorized with codes designed to capture the key aspects of the Cold Rush with a time/space connection, and the coded photos form now a Cold Rush photo archive.

We curated thirteen Cold Rush Science postcards with Alexandre Duchêne from this archive. We aimed to create mobile material objects that would carry an evocative visual vignette of the Cold Rush and convey some of the project's key results. We selected pictures that captured the various aspects of the Cold Rush but, together, created a sense of how it looked alike in the Finnish Arctic. The photos were accompanied by captions printed on the back of the postcards. They indexed the particular key finding at stake, but deliberately in a somewhat ambiguous way to appeal to the interlocutors' curiosity and leave space for them to figure out the connections between the text, picture, and the Cold Rush. To enhance the postcard genre and its function as a mobile object, we printed address lines at the backside of the cards. Together, the photo, the caption, and the postcard genre produce, in Deleusian terms, molecular assemblages that function at least in two ways: they deterritorialize the postcard genre from the tourism terrain and reterritorialized them into scientific knowledge production territory, and they deterritorialize scientific knowledge from academic genres to a more informal genre of postcards. These movements open the possibility of discovering and discussing the Cold Rush findings and their connections across different audiences, including stakeholders, locals, students, and colleagues. To facilitate these possibilities, we send out these postcards to our colleagues. I have used them in my conversations with research participants, locals, stakeholders, colleagues, and students. To facilitate more accidental connections, I have also placed the postcards in the key sites where we conducted our fieldwork so anyone can pick them up. Every now and then, I return to these hotspots and refill the stack.

To facilitate rhizomatic reading, I will introduce the key participants at the beginning of each empirical chapter. I find anonymizing places and persons to be a nuanced process, perhaps best described as a negotiation. Even after closely following the institutional and academic ethical guidelines, like those of the American Anthropology Association or the Finnish Science Association, many choices remain. I have anonymized the identifying information regarding the participants, but, still, given the smallness of the communities, there is a possibility that somebody who knows local conditions can speculate, and this aspect was discussed with and approved by the participants. They have also read versions of this book and approved how the data related to them is used and presented in this book. I use the actual names of the places across four municipalities where I have worked when it is relevant for understanding the historical processes leading to the current situation. This is particularly the case with Chapter 3, where I trace the intertwined relationships between mining, colonialism, and capitalism, all connected to the Tornio River and the first mine by this river.

References

Ahlness, Ellen. 2019. 'Incompatible Futures: Frontier Nostalgia and Southern Discourses of the Arctic'. *Arctic Yearbook* 2019 (November): 1–20.

Allan, Kori. 2019. 'Volunteering as Hope Labour: The Potential Value of Unpaid Work Experience for the Un- and Under-Employed'. *Culture, Theory and Critique* 60 (1): 66–83. https://doi.org/10.1080/14735784.2018.1548300.

Andersson, Daniel. 2020. 'Indigenous Place-Names in (Post)colonial Contexts: The Case of Ubmeje in Northern Sweden'. *Scandinavian Studies* 92 (1): 104–26.

Bazzul, Jesse, and Sara Tolbert. 2017. 'Reassembling the Natural and Social Commons'. In *Educational Commons in Theory and Practice: Global Pedagogy and Politics*, edited by Alexander J. Means, Derek R. Ford, and Graham B. Slater, 55–73. New York: Palgrave Macmillan US. https://doi.org/10.1057/978-1-137-58641-4_4.

Bear, Laura. 2020. 'Speculation: A Political Economy of Technologies of Imagination'. *Economy and Society* 49 (1): 1–15. https://doi.org/10.1080/03085147.2020.1715604.

Bebbington, Anthony, Nicholas Cuba, and John Rogan. 2014. 'The Overlapping Geographies of Resource Extraction'. *ReVista (Cambridge)* 13 (2): 20–23.

Bell, Lindsay A. 2023. *Under Pressure: Diamond Mining and Everyday Life in Northern Canada.* Toronto: University of Toronto Press. https://doi.org/10.3138/9781487548575.

Bennett, Jane. 2010. *Vibrant Matter: A Political Ecology of Things.* Durham: Duke University Press. https://doi.org/10.1515/9780822391623.

Bignall, Simone, and Paul Patton, eds. 2010. *Deleuze and the Postcolonial. Deleuze Connections.* Edinburgh: Edinburgh University Press.

Braidotti, Rosi. 2019. 'A Theoretical Framework for the Critical Posthumanities'. *Theory, Culture & Society* 36 (6): 31–61. https://doi.org/10.1177/0263276418771486.

Bryant, Rebecca, and Daniel M. Knight. 2019. *The Anthropology of the Future. New Departures in Anthropology*. Cambridge: Cambridge University Press. https://doi.org/10.1017/9781108378277.

Buchanan, Ian. 2021. *Assemblage Theory and Method*. London: Bloomsbury Publishing Plc. https://doi.org/10.5040/9781350015579.

Business Lapland | Opportunities Above Ordinary. n.d. House of Lapland. https://www.lapland.fi/business/. Accessed 11 August 2023.

Carrasco, Sebastián, Javier Hernández, and Valentina Cariaga. 2023. 'The Temporalities of Natural Resources Extraction: Imagined Futures and the Spatialization of the Lithium Industry in Chile'. *The Extractive Industries and Society* 15 (September): 101310. https://doi.org/10.1016/j.exis.2023.101310.

Chakma, Urmee, and Shaila Sultana. 2023. 'Colonial Governmentality and Bangladeshis in the Anthropocene: Loss of Language, Land, Knowledge, and Identity of the Chakma in the Ecology of the Chittagong Hill Tracts in Bangladesh'. *Ethnicities*, November, 14687968231219521. https://doi.org/10.1177/14687968231219521.

Chakrabarty, Dipesh. 2018. 'Anthropocene Time'. *History and Theory* 57 (1): 5–32. https://doi.org/10.1111/hith.12044.

Clarke, Bryan, and Jim Parsons. 2013. 'Becoming Rhizome Researchers'. *Reconceptualizing Educational Research Methodology* 4 (1): 35-43. https://doi.org/10.7577/rerm.685.

Cornips, Leonie, and Louis van den Hengel. 2021. 'Place-Making by Cows in an Intensive Dairy Farm: A Sociolinguistic Approach to Nonhuman Animal Agency'. In *Animals in Our Midst*, edited by Bernice Bovenkerk and Jozef Keulartz, 177–201. The International Library of Environmental, Agricultural and Food Ethics. Berlin: Springer. https://doi.org/10.1007/978-3-030-63523-7_11.

Coulthard, Glen Sean. 2014. *Red Skin, White Masks: Rejecting the Colonial Politics of Recognition*. Indigenous Americas. Minneapolis: University of Minnesota Press.

D'Angelo, Lorenzo, and Robert J. Pijpers. 2018. 'Mining Temporalities: An Overview'. *The Extractive Industries and Society* 5 (2): 215–22. https://doi.org/10.1016/j.exis.2018.02.005.

Deleuze, Gilles. 1986/1988. *Foucault*. Minneapolis: University of Minnesota Press.

Deleuze, Gilles, and Félix Guattari. 1972/1977. *Anti-Oedipus: Capitalism and Schizophrenia*. Translated by Robert Hurley, Mark Seem, and Helen R. Lane. New York: Viving Press.

———. 1980/1987. *A Thousand Plateaus: Capitalism and Schizophenia*. Translated by Brian Massuni. London: Bloomsbury Academic.

Deleuze, Gilles, and Michel Foucault. 1977. 'Intellectuals and Power'. In *Language, Counter-Memory, Practice: Selected Essays and Interviews*. Edited by Donald F. Bouchard, 205–17. Ithica: Cornell University Press.

Demuth, Bathsheba. 2020. *Floating Coast: An Environmental History of the Bering Strait*. New York, NY: W.W. Norton & Company.

Dosse, François. 2010. *Gilles Deleuze and Félix Guattari: Intersecting Lives. European Perspectives*. New York: Columbia University Press.

———. 2016. 'Deleuze and Foucault: A Philosophical Friendship'. In *Between Deleuze and Foucault,* edited by Nicolae Morar, Nicolae, Thomas Nail, and Daniel W. Smith, 11–37. Edinburgh: Edinburgh University Press. https://doi.org/10.1515/9781474415095.

Ferry, Elizabeth Emma. 2013. *Minerals, Collecting, and Value across the U.S.-Mexico Border. Tracking Globalization*. Bloomington: Indiana University press.

Ferry, Elizabeth. 2020. 'Speculative Substance: "Physical Gold" in Finance'. *Economy and Society* 49 (1): 92–115. https://doi.org/10.1080/03085147.2019.1690254.

Foucault, Michel. 1967/1972. *Archaeology of Knowledge*. Translated by A.M. Sheridan Smith. New York: Pantheon Books.

———. 1972/1977. *Discipline and Punish: The Birth of the Prison*. Translated by A.M. Sheridan Smith. New York: Pantheon Books.

———. 1977/1984. 'Nietzsche, Genealogy, History'. In *The Foucault Reader*, edited by Paul Rabinow, 76–100. New York: Pantheon.

Glissant, Édouard. 1997. *Poetics of Relation*. Michigan: University of Michigan Press. https://press.umich.edu/Books/P/Poetics-of-Relation.

Graugaard, Naja Dyrendom. 2019. "Without Seals, There Are No Greenlanders": Colonial and Postcolonial Narratives of Sustainability and Inuit Seal Hunting'. In *The Politics of Sustainability in the Arctic*, edited Ulrik Pram Gad and Jeppe Strandsbjerg, 74–93. London: Routledge.

Gros, Frédéric, and Samantha Bankston. 2016. 'Deleuze's Foucault: A Metaphysical Fiction'. In *Between Deleuze and Foucault*, edited by Nicolae Morar, Thomas Nail, and Daniel W. Smith, 128–38. Edinburgh: Edinburgh University Press. https://doi.org/10.1515/9781474415095

Halvaksz, Jamon Alex. 2008. 'Whose Closure? Appearances, Temporality, and Mineral Extraction in Papua New Guinea'. *Journal of the Royal Anthropological Institute* 14 (1): 21–37. https://doi.org/10.1111/j.1467-9655.2007.00476.x.

Haraway, Donna. 2015. 'Anthropocene, Capitalocene, Plantationocene, Chthulucene: Making Kin'. *Environmental Humanities* 6 (1): 159–65. https://doi.org/10.1215/22011919-3615934.

Heller, Monica, Sari Pietikäinen, and Joan Pujolar. 2018. *Critical Sociolinguistic Research Methods: Studying Language Issues That Matter*. 1st ed. New York: Routledge.

Hong, Sun-ha. 2020. *Technologies of Speculation: The Limits of Knowledge in a Data-Driven Society*. New York: New York University Press.

Hovens, Daan. 2023. 'Breakdowns and Assemblages: Including Machine-Actants in Sociolinguistic Ethnographies of Blue-Collar Work Environments'. *Journal of Sociolinguistics* 27 (1): 3–23. https://doi.org/10.1111/josl.12565.

Jokinen, Mikko. 2019. 'Lapin ympäristökiistojen kulttuuriset tekijät'. *Dissertationes Forestales* 2019 (281). Finnish Society of Forest Science. https://www.dissertationesforestales.fi/article/10231.

Joona, Tanja L., and Juha Joona. 2023. 'Natural Resource Development in the Arctic and the Question of Saami Land Rights in Finland'. In *The Routledge Handbook of Polar Law*, edited by Rachael Johnstone, Yoshifumi Tanaka, and Vibe Ulfbeck, 437–55. London: Routledge. https://doi.org/10.4324/9781003404828-33.

Junka-Aikio, Laura. 2023. 'Whose Settler Colonial State? Arctic Railway, State Transformation and Settler Self-Indigenization in Northern Finland'. *Postcolonial Studies* 26 (2): 279–301. https://doi.org/10.1080/13688790.2022.2096716.

Kelly-Holmes, Helen. 2016. 'Theorising the Market in Sociolinguistics'. In *Sociolinguistics: Theoretical Debates*, edited by Nikolas Coupland, 157–72. Cambridge: Cambridge University Press. https://doi.org/10.1017/CBO9781107449787.008.

Kerslake, Christian. 2015. 'Marxism and Money in Deleuze and Guattari's Capitalism and Schitzofrenia: On the Conflict between the Theories of Suzanne De Brunhoff and Berbard Schmitt'. *Parrhesia*, no. 22, 38–78.

Keskitalo, Eva Carina H., ed. 2019. *The Politics of Arctic Resources: Change and Continuity in the 'Old North' of Northern Europe*. New York: Routledge.

Kröger, Markus. 2016. 'Spatial Causalities in Resource Rushes: Notes from the Finnish Mining Boom'. *Journal of Agrarian Change* 16 (4): 543–70. https://doi.org/10.1111/joac.12113.

Kuokkanen, Rauna. 2019. 'At the Intersection of Arctic Indigenous Governance and Extractive Industries: A Survey of Three Cases'. *The Extractive*

Industries and Society 6 (1): 15–21. https://doi.org/10.1016/j.exis.2018.08.011.

———. 2022. 'All I See Is White: The Colonial Problem in Finland'. In *Finnishness, Whiteness and Coloniality*, edited by Josephine Hoegaerts, Tuire Liimatainen, Laura Hekanaho, and Elizabeth Peterson, 291–314. Helsinki University Press. https://doi.org/10.33134/HUP-17-12.

———. 2023. 'From Indigenous Private Property to Full Dispossession—The Peculiar Case of Sápmi'. *Comparative Legal History* 11 (1): 23–44. https://doi.org/10.1080/2049677X.2023.2207380.

Lamb, Gavin. 2024. *Multispecies Discourse Analysis: The Nexus of Discourse and Practice in Sea Turtle Tourism and Conservation.* London: Bloomsbury Publishing Plc.

Lamb, Gavin, and Christina Higgins. 2020. 'Posthumanism and Its Implications for Discourse Studies'. In *The Cambridge Handbook of Discourse Studies*, edited by Alexandra Georgakopoulou and Anna De Fina, 350–70. Cambridge Handbooks in Language and Linguistics. Cambridge: Cambridge University Press. https://doi.org/10.1017/9781108348195.017.

Lapinluotsi 2022. Accessed April 15, 2014. www.lapinluotsi.fi.

Larsen, Joan Nymand, and Lee Huskey. 2015. 'The Arctic Economy in a Global Context'. In *The New Arctic*, edited by Birgitta Evengård, Joan Nymand Larsen, and Øyvind Paasche, 159–74. London: Palgrave Macmillan. https://doi.org/10.1007/978-3-319-17602-4_12.

Lassila, Maija M. 2018. 'Mapping Mineral Resources in a Living Land: Sami Mining Resistance in Ohcejohka, Northern Finland'. *Geoforum* 96 (November): 1–9. https://doi.org/10.1016/j.geoforum.2018.07.004.

Legg, Stephen. 2011. 'Assemblage/Apparatus: Using Deleuze and Foucault'. *Area* 43 (2): 128–33. https://doi.org/10.1111/j.1475-4762.2011.01010.x.

Lehtola, Veli-Pekka. 1997. *Rajamaan identiteetti: Lappilaisuuden rakentuminen 1920- ja 1930-luvun kirjallisuudessa.* Suomalaisen Kirjallisuuden Seuran toimituksia 665. Helsinki: Suomalaisen Kirjallisuuden Seura.

———. 2002 'The Saami Siida and the Nordic States from the Middle Ages to the Beginning of the 1900s'. In *Conflict and Cooperation in the North*, edited by Kristiina Karppi and Johan Eriksson. Umeå: Umeå Universitet.

———. 2012. *Saamelaiset suomalaiset: kohtaamisia 1896–1953.* Suomalaisen Kirjallisuuden Seuran toimituksia 1351. Helsinki: Suomalaisen Kirjallisuuden Seura.

———. 2021. 'Contested Sámi Histories in Finland'. In *Sámi Research in Transition*, edited by Laura Junka-Aikio, Jukka Nyyssönen and Veli-Pekka Lehtola. London: Routledge. https://doi.org/10.4324/9781003090830.

———. 2022. *Entiset elävät meissä: Saamelaisten historiat ja Suomi*. Helsinki: Gaudeamus.

Li, Zi-Chen, Wen-Bin Sun, Cai-Xia Liang, Xu-Huang Xing, and Qing-Xiang Li. 2023. 'Arctic Warming Trends and Their Uncertainties Based on Surface Temperature Reconstruction under Different Sea Ice Extent Scenarios'. *Advances in Climate Change Research* 14 (3): 335–46. https://doi.org/10.1016/j.accre.2023.06.003.

Maegaard, Marie, and Kristine Køhler Mortensen. 2022. 'Colonial Labels and the Imagined Innocence of Past Times: Debating Language and Spatial Representations of the Danish/Greenlandic Relation'. *Language in Society* 51 (5): 775–95. https://doi.org/10.1017/S0047404522000604.

Mahmood, Saba. 2012. *Politics of Piety: The Islamic Revival and the Feminist Subject*. Princeton: Princeton University Press.

May, Todd. 2005. *Gilles Deleuze: An Introduction*. New York: Cambridge University Press.

Morar, Nicolae, Thomas Nail, and Daniel W. Smith. 2016. Eds. *Between Deleuze and Foucault*. Edinburg: Edinburgh University Press.

Narotzky, Susana, and Niko Besnier. 2014. 'Crisis, Value, and Hope: Rethinking the Economy: An Introduction to Supplement 9'. *Current Anthropology* 55 (S9): S4–16. https://doi.org/10.1086/676327.

Nichols, Robert. 2010. 'Postcolonial Studies and the Discourse of Foucault: Survey of a Field of Problematization'. *Foucault Studies*, September, 111–44. https://doi.org/10.22439/fs.v0i9.3062.

Nuttall, Mark. 2018. 'Arctic Weather Words'. *Anthropology News* 59 (2): e37–43. https://doi.org/10.1111/AN.793.

Nuttall, Mark, Torben R. Christensen, and Martin J. Siegert, eds. 2018. *The Routledge Handbook of the Polar Regions*. London: Routledge.

Ojala, Anna-Liisa. 2020. 'Being an Athlete and Being a Young Person: Technologies of the Self in Managing an Athletic Career in Youth Ice Hockey in Finland'. *International Review for the Sociology of Sport* 55 (3): 310–26. https://doi.org/10.1177/1012690218801303.

Outakoski, Hanna. 2022. 'It May Be Invisible to You but It Still Affects Us: Extending the Comprehensive View on Language and Writing'. In *Indigenising Education and Citizenship*, edited by Trojer A Olsen and Hilde Sollid, 214–36. Books. Oslo: Universitetsforlaget. https://doi.org/10.18261/9788215053417-2022-12.

Paajanen, Annukka Maaria. 2024. 'Seeds in Translation: From Multispecies Environments into Frozen Assets'. *Interconnections* 3 (1): 20–38.

Peebles, Gustav. 2010. 'The Anthropology of Credit and Debt'. *Annual Review of Anthropology* 39 (1): 225–40. https://doi.org/10.1146/annurev-anthro-090109-133856.

Pennycook, Alastair. 2016. 'Mobile Times, Mobile Terms: The Trans-Super-Poly-Metro Movement'. In Sociolinguistics: Theoretical Debates, edited by Nikolas Coupland, 201–16. Cambridge: Cambridge University Press.

Pietikäinen, Sari. 2016. 'Critical Debates: Discourse, Boundaries and Social Change'. In *Sociolinguistics: Theoretical Debates*, edited by Nikolas Coupland, 263–81. Cambridge: Cambridge University Press.

———. 2019. 'Discussion: Language in Nature Resource Economies'. *International Journal of the Sociology of Language* 2019 (258): 171–76. https://doi.org/10.1515/ijsl-2019-2033.

———. 2021. 'Powered by Assemblage: Language for Multiplicity'. *International Journal of the Sociology of Language* 2021 (267–268): 235–40. https://doi.org/10.1515/ijsl-2020-0074.

Pietikäinen, Sari, Helen Kelly-Holmes, Alexandra M. Jaffe, and Nikolas Coupland. 2016. *Sociolinguistics from the Periphery: Small Languages in New Circumstances*. Cambridge: Cambridge University Press.

Pietikäinen, Sari, and Kori Allan. 2021. 'Jobs for Life?': Mining Temporalities in a Transforming Arctic Periphery'. In *Language, Global Mobilities, Blue-Collar Workers and Blue-Collar Workplaces*, edited by Kellie Gonçalves and Helen Kelly-Holmes, 27–48. London: Routledge. https://doi.org/10.4324/9780429298622-2.

Pietikäinen, Sari, and Christine Hegel. 2021. 'Assembling Success: Desire, Worker Mobility, and Value Creation in Finnish Ice Hockey'. *Anthropology of Work Review* 42 (2): 71–81. https://doi.org/10.1111/awr.12221.

Pietikäinen, Sari, and Helen Kelly-Holmes. 2013. Eds. *Multilingualism and the Periphery*. Oxford: Oxford University Press.

Pietikäinen, Sari, and Anne Mäntynen. 2019. *Uusi kurssi kohti diskurssia*. Tampere: Vastapaino.

Povinelli, Elizabeth A. 2011. 'THE GOVERNANCE OF THE PRIOR'. *Interventions* 13 (1): 13–30. https://doi.org/10.1080/1369801X.2011.545575.

Pratt, Mary Louise. 1991. 'Arts of the Contact Zone'. *Profession*, 33–40.

Rantanen, Mika, Alexey Yu Karpechko, Antti Lipponen, Kalle Nordling, Otto Hyvärinen, Kimmo Ruosteenoja, Timo Vihma, and Ari Laaksonen. 2022. 'The Arctic Has Warmed Nearly Four Times Faster than the Globe since 1979'. *Communications Earth & Environment* 3 (1): 1–10. https://doi.org/10.1038/s43247-022-00498-3.

Rantanen, Pekka, and Jarno Valkonen. 2011. 'Ulkomaalaiset metsämarjapoimijat Suomessa'. Rovaniemi: Lapin yliopisto.
Ruddick, Susan. 2010. 'The Politics of Affect: Spinoza in the Work of Negri and Deleuze'. *Theory, Culture & Society* 27 (4): 21–45. https://doi.org/10.1177/0263276410372235.
Said, Edward. 1995. 'Secular Interpretation, the Geographical Element, and the Methodology of Imperialism'. In *After Colonialism. Imperial histories and Postcolonial Displacements,* edited by Gyan Prakash, *21–39.* Princeton: Princeton University Press. https://doi.org/10.1515/9781400821440.21.
Sarivaara, Erika Katjaana. 2012. *Statuksettomat saamelaiset. Paikantumisia saamelaisuuden rajoilla.* Doctoral thesis. Sámi allaskuvla. http://hdl.handle.net/11250/177093
Schweitzer, Peter. 2022. 'Between Resource Frontier and Self-Determination: Colonial and Postcolonial Developments in the Arctic'. In *Global Arctic: An Introduction to the Multifaceted Dynamics of the Arctic*, edited by Matthias Finger and Gunnar Rekvig, 351–67. London: Palgrave Macmillian. https://doi.org/10.1007/978-3-030-81253-9_18.
Scollon, Ronald, and Suzanne B. K. Scollon. 2004. *Nexus Analysis: Discourse and the Emerging Internet*. London: Routledge.
Shankar, Shalini, and Jillian R. Cavanaugh. 2012. 'Language and Materiality in Global Capitalism'. *Annual Review of Anthropology* 41 (1): 355–69. https://doi.org/10.1146/annurev-anthro-092611-145811.
Simpson, Audra. 2017. 'The Ruse of Consent and the Anatomy of "Refusal": Cases from Indigenous North America and Australia'. *Postcolonial Studies* 20 (1): 18–33. https://doi.org/10.1080/13688790.2017.1334283.
Sollid, Hilde, and Torjer Olsen. 2019. 'Indigenising Education: Scales, Interfaces and Acts of Citizenship in Sápmi'. *Junctures: The Journal for Thematic Dialogue*, no. 20 (November). https://doi.org/10.34074/junc.20029.
Spangen, M., A.-K. Salmi, T. Aikas, and V.-P. Lehtola. 2015. 'Sami Histories, Colonialism, and Finland'. *Arctic Anthropology* 52 (2): 22–36. https://doi.org/10.3368/aa.52.2.22.
Spivak, Gayatri. 1988. 'Can the Subaltern Speak?' In *Marxism and the Interpretation of Culture,* edited by Cary Nelson and Lawrence Grossberg, 271–313. Illinois: University of Illinois Press.
Strömmer, Maiju. 2021. 'In the Name of Security: Governmentality Apparatus in a Multilingual Mine in Arctic Finland'. *Journal of Sociolinguistics* 25 (2): 217–34. https://doi.org/10.1111/josl.12458.
Schweitzer, Peter. 2022. 'Between Resource Frontier and Self-Determination: Colonial and Postcolonial Developments in the Arctic'. In *Global Arctic:*

An Introduction to the Multifaceted Dynamics of the Arctic, edited by Matthias Finger and Gunnar Rekvig, 351–67. Cham: Springer International Publishing. https://doi.org/10.1007/978-3-030-81253-9_18.

Tervaniemi, Saara, and Päivi Magga. 2018. 'Belonging to Sápmi – Sámi Conceptions of Home and Home Region'. In *Knowing from the Indigenous North*, edited by Thomas Hylland Eriksen, Sanna Valkonen and Jarno Valkonen, 75–90. London: Routledge.

The Economy of the North 2020 ECONOR. 2021. Oslo-Kongsvinger: Statistisk sentralbyrå.

Tsing, Anna Lowenhaupt. 2015. *The Mushroom at the End of the World: On the Possibility of Life in Capitalist Ruins*. Princeton: Princeton University press.

Tsing, Anna, Heather Swanson, Elaine Gan, and Nils Bubandt. 2017. *Arts of Living on a Damaged Planet: Ghosts and Monsters of the Anthropocene*. University of Minnesota Press. https://www.jstor.org/stable/https://doi.org/10.5749/j.ctt1qft070.

Valkonen, Sanna. 2009. *Poliittinen Saamelaisuus*. Tampere: Vastapaino.

———. 2018. 'Conceptual Governance on Defining Indigeneity—The Sámi Debate in Finland'. In *Knowing from the Indigenous North*, edited by Thomas Hylland Eriksen, Sanna Valkonen and Jarno Valkonen, 143–162. London: Routledge.

Valkonen, Sanna, Áile Aikio, Saara Alakorva, and Sigga-Marja Magga. 2022. *The Sámi World*. London: Routledge. https://doi.org/10.4324/9781003025511.

Valkonen, Sanna, and Jarno Valkonen. 2017. 'The Non-State Sámi: Struggle for Indigenous Citizenship in the European North'. In *Shaping Citizenship: A Political Concept in Theory, Debate and Practice*, edited by Claudia Wiesner, Anna Björk, Hanna-Mari Kivistö and Katja Mäkinen, 138-52. London: Routledge.

Valkonen, Sanna, Jarno Valkonen, and Veli-Pekka Lehtola. 2017. 'An Ontological Politics of and for the Sámi Cultural Heritage—Reflections on Belonging to the Sámi Community and the Land'. In *Indigenous Peoples' Cultural Heritage*, edited by Sanna Valkonen, Alexandra Xanthaki, Leena Heinämäki, and Piia Nuorgam, 149–74. Leiden: Brill. https://www.jstor.org/stable/10.1163/j.ctv2gjwsw2.10.

2

Researching Cold Rush: Critical Assemblage Analysis

2.1 Connecting the Hotspots

Fig. 2.1 Arctic blue moment

S. Pietikäinen, *Cold Rush*, Language and Globalization,
https://doi.org/10.1007/978-3-031-63995-1_2

In 2005, the last mainly Northern Sámi-language primary school in the Finnish Arctic closed. Politically, the decision was a difficult one, since the school was one of the region's few remaining predominantly Sámi language schools. Historically, the school, its students, and teachers were valued as the heartland of the Sámi political movement and language revitalization activism. In 2004, when I conducted an ethnography on multilingualism at the school, it was already clear to the students and teachers that, despite resistance from the community, the school would eventually close, since there were not enough pupils to maintain it. The area's ageing population and the shift from primary sector work to services-sector jobs materialized as a diminishing number of families with school-age children. Since the 1970s, unemployment in Finland has been highest in Lapland, and people have left for the southern urban centres or abroad, mainly to Sweden. No jobs, no families, no children, no village school.

The school building was closed for years before it moved from the domain of a state education service to the domain of the Arctic nature tourism business. An entrepreneur bought the building and converted the school into a remote nature hotel. A few years later, new entrepreneurs bought the school/hotel building and created a new brand: "silence, the luxury of nature", as the slogan says. The silence brand transformed the vast, uninhabited nature surrounding the school/hotel and the lack of services and infrastructure into a relaxing nature experience away from the fast tempo and demands of normal life (Fig. 2.1).

Changing the value of silence from a lack of buzz, business, and infrastructure into luxury, tranquillity, and serenity requires changes in material, discursive, and affective frameworks surrounding emptiness. Silence needs to be assembled with new elements to re-value and reterritorialize it within the terrain of nature tourism. Silence, as a product, must connect with the right conditions, including logics, sustainability, and, increasingly, security. It must be organized in a particular way for it to make sense to tourists, on the one hand, and create profit for entrepreneurs, on the other hand. Narratives about the role of silence and its value in the region are embedded in the familiar tourism genres, such as waiting, photographing, and gazing at the wonders of nature, facilitating tourists' participation and expectations, and leading, in the best conditions, to feelings of luxury and relaxation. Under the right conditions, the new assemblage consisting of the school/hotel building, surrounding nature, stories about the region, and lack of urban infrastructure produces silence as a luxury product.

To examine the revaluing of Arctic natural resources, I have developed a framework called *critical assemblage analysis*, in which I connect three distinct yet interrelated concepts: Foucauldian understanding of discourse, Deleuzian-Guattarian rhizomatic ontology, and Deleuzian-Guattarian concept of assemblage, and I apply them in examining the Cold Rush with the help of ethnography. Using this framework, I examine the Cold Rush moments when codes, relationships, and functions of the Arctic natural resources are reconfigured, resulting in rearrangements in their meaning and value as well as in the relationships around them. These moments do not occur in a vacuum, but they unfold over time and space and can be examined as situated interactions between material and discursive elements embedded in historical processes and power dynamics. The moments of revaluing are connected to processes of de- and reterritorialization, which Deleuze and Guattari (1980/1987) identify as key capitalist processes. Sometimes, revaluing Arctic natural resources works, but not always or as expected. Some value transformations are enduring, others are only ephemeral, and a few remain speculative. The framework examines these Cold Rush multiplicities in order to understand how Arctic natural resources become revalued into good jobs, for example, or superberries, and lucrative Northern Lights businesses. In the following, I will unpack the concepts of rhizome, assemblage, and discourse as I use them to examine the revaluing of Arctic natural resources and how I connect them in a critical assemblage analysis of the Cold Rush. While the order between these concepts is interchangeable, in what follows, I will discuss them by working from the broader, ontological orientation towards more analytical concepts, since this is how I have used them in examining the Cold Rush. Thus, I start with rhizomatic ontology and move on to the concepts of assemblage and discourse before introducing the analytical framework, critical assemblage analysis.

2.2 Rhizomatic Rush: Entangled Connections

Fig. 2.2 Rhizome on ice

The value of natural resources is often discussed with the help of mutually exclusive binaries, such as extraction-preservation, local-global, pure-polluted, and protection-profit. But this is only one possibility; rhizomatic ontology, developed by Deleuze and Guattari's (1980/1987), offers an alternative. The concept is borrowed from botany where it refers to the vertiginous growth potential of the weed and subterranean plant stem, which sends out roots and shoots from its nodes. Importantly, the rhizome does not have one organising core, centre, or hierarchy, but it is dynamic, multidirectional, and evolving. It is a web of connections rather than an intersection between two options. The rhizome has multiple entry and exit points, and it can expand into new directions and grow new roots to replace broken ones. It has no beginning, no end, no particular shape, but rather, it is in the middle, in process, evolving. Importantly, it escapes determinism, and the loop of fixed relations, structures, and categories and, in this way, offers a way forward from essentializing thinking. The direction and logic of a rhizome come from its function, not a pre-determined, unchanging structure or set of relations. In Deleuze and Guattari's use of the term, the rhizome maps a networked, relational, and transversal thought process (Deleuze and Guattari 1980/1987: 12; see also Coleman and Ringrose 2013) (Fig. 2.2).

The rhizome is hard to think of in isolation, and it is often discussed in relation to its opposite, a tree structure. In the work by Deleuze and Guattari (1980/1987), the rhizome builds on connections, while the tree structure employs hierarchical organization (May 2005: 133–134). As Deleuze and Guattari famously wrote:

> The tree is affiliations but the rhizome is alliances, unique alliances. The tree imposes the verb "to be" but the fabric of the rhizome is the conjunctions "and…and…and"… (1980/1987: 25)

The tree model has a long tradition in dominant Western political thought and is typically applied in state governance, nation-building, and political projects. Also, biological species or languages are organized according to a tree model (Adkins 2015: 23). A tree model builds on clear, fixed, relatively stable hierarchies, binaries, and categories. It draws on one dominant story about the historical development and future directions. The ambivalences and tensions around the boundaries, shifts, and interruptions are disregarded, silenced, or simply ignored (Nail 2017). In the Cold Rush, the power of the tree model lies in its capacity to simplify the multiplicity of how natural resources are governed, profited from, and protected. It masks the differences between people's various ways of identifying with natural resources, unites people around one category or cause, and sets aside people's own experiences that may conflict with the totality of the tree model.

The rhizome model is typically applied in counter-politics, resistance art, and radical political spaces and projects. It challenges essentialism, logocentrism, and fixed modern subjects and categories as its planar movement resists chronology and hierarchical organization, favouring, instead, a nomadic system of growth, unpredictability, and creativity (Gehman et al. 2022: 293–204). The rhizome model provides a critique of dichotomous representations of fixed boundaries, unchanging categories, and totalizing politics, which may now be inadequate to capture and explain what happens on the ground. Instead, rhizome focuses on connections and multiplicities that enable us to see productive gaps and competing forms of knowledge that play out in practices of mobility and patterns of circulation across nation-state imaginaries and scalar visions (Wallin 2010: 83).

A reason to apply a rhizome model in research is to find a way to account for the multiplicities encountered in the field—how processes, changes, and practices interact in emergent ways (Adkins 2015; Appadurai 1996). Rhizomatic ontology provides an alternative mode for thinking, knowing, and researching by insisting on retaining multiplicity, the process of becoming, and interactions in various dimensions (Nail 2017). Rhizome lends a metaphoric physicality to complex processes, enabling us to see connections and avoid subsuming them under one focus or category (Adkins 2015: 25). Deleuze and Guattari (1980/1987) illustrate the idea of rhizome with the help of cartography: the rhizome is a map that allows us to notice potential routes, entry points, and crossroads. In examining the Cold Rush, the rhizomatic approach has the capacity to circumvent the trap of a priori categorizations and instead encounter the ongoing and overlapping processes of natural resource politics and economics. It provides, for me, a way to look at both stability and change at a moment when previous categories and links between the Arctic, its resources, and the categories and rationales traditionally used to govern and value them are being debated and reconfigured. People and institutions need to figure out how to proceed, specifically, whether to defend existing categories and rationalities, update them or construct novel ones that will work for them in new circumstances.

Both tree and rhizome thinking have their value. The tree model works in close alliance with the cultural, political, and economic power centres, whereas the rhizomatic model connects with multiple experiences of people, activism, and unfinished, ongoing political, and economic processes. The tree model produces causal connections and linear trajectories, while the rhizomatic model fosters connections, possibilities, creativity, and discovery. Deleuze and Guattari (1980/1987) caution against a celebration of either model. Rather, they suggest that we can work with the premise that neither the rhizome nor the tree model is primary, but they may vary, change, and intertwined, depending on the function and the situation at hand.

In examining the Cold Rush, I use the concept of rhizome to study entanglements in the revaluing of natural resources over time and place under the changing conditions of the Finnish Arctic. My starting point is the "knotting" of rhizomatic and hierarchical revaluing processes: this requires mapping the trajectories and circulations of a particular natural resource as well as understanding the conditions and consequences that produce a particular value, as well as how it matters to people. Rhizomatic thinking allows me to search for multiple connections from various perspectives and experiences that are not rooted in a single moment or event. It provides me with an alternative way to think and conceptualize multiplicity in relation to the revaluing of natural resources.

2.3 Assemblage Effect: From Essence to Productive Interactions

Fig. 2.3 Star assemblages

At first glance, the indefinite stars shimming in the night sky may look like a silent sea of flickering lights. But once you begin to look for familiar star constellations—connections between stars creating Big Dipper, Orion, Gemini—the stars and the lines between them, you and your location begin to have particular functions. Humans use star constellations for navigation and geographical orientation, to mark the passage of time and define the calendar, as well as for cultural and religious practices. Some other animals, like seals, certain birds, and beetles, for example, use star constellations to navigate and find food and mates. This is the core idea of assemblage: a group of stars becomes more than the sum of its parts through a productive interaction between stars, earth, and living beings. Assemblage is not any kind of random or complex collection of things; it emerges when a function and interaction between the elements emerge (Buchanan 2015, 2021; Nail 2017) (Fig. 2.3).

As outlined in the previous chapter, Deleuze and Guattari define assemblage as an arrangement of heterogeneous elements powered by desire and influenced by capitalist transformations (Deleuze and Guattari 1980/1987; Hamalakis 2017). At its most basic, an assemblage combines material (actions, bodies, things) and discursive (expressions, ideas, affects) elements (Buchanan 2021: 33, see also Feely 2020; Dovey 2020). This core idea stems from the theory of content-form dynamics presented by a Danish linguist Hjelmslev (1899–1965). Deleuze and Guattari (1980/1987) expand on this idea by emphasizing the interactions between various elements, resulting in dissolving the persisting opposition inherent in binary thinking in relation to material/discursive, micro/macro, mechanic/expressive, technological/societal, and human/nonhuman.

Assemblage, originating from the French *agencement*, refers to both the act of assembling diverse elements—a berry, the Arctic, a health discourse—and their arrangement for a specific purpose, such as a superberry for international health markets. Many have written about the difficulties of productively translating the multiple meanings of the French word *agencement* as used by Deleuze and Guattari, and while some recommend terms such as organization, arrangement, or layout, the term assemblage is most commonly used (Buchanan 2021: 19–21). Gehman et al. (2022) suggest that using “assembling” would better capture the original meaning of the concept, and Buchanan's (2021: 21) preference would be "working arrangement" to underline its nature as an ongoing process. While there is no perfect match, these translation tensions underline the multiplicity of processes embedded in assemblage: it entails both assembling and arranging processes, ongoing processes, or re-assembling.

Importantly, it is the interaction between elements that enables an assemblage to become more than the sum of its parts. Deleuze and Guattari (1980/1987: 58) give an example of a horse, a body, and an arrow: only their interaction turns them into a weapon. Similarly, simply putting a body, skates, a stick, and a puck on the ice rink does not make a professional hockey player: productive interaction between a set of material and discursive elements is needed (Pietikäinen and Hegel 2021). Furthermore, the elements must be brought together under particular conditions, made to function in specific ways to accomplish their conjoined capacities and goals (Wise 2005): the hockey player needs their team, audience, and competition to function as a professional athlete.

In researching the Cold Rush, I wanted to engage with assemblage thinking as it provides an alternative to the logic of fixed essences (Nail 2017). Assemblage takes interactions, relations, heterogeneity, and connections as a starting point and offers one way of dealing with the dynamics of contingency and structure, organization and change (Wise 2005: 77). Through the focus on interaction, it emphasizes fluidity and multiple functionalities through elements and their connectivity. Critically, the relationships of component parts are not eternally stable and fixed; instead, they can be displaced and replaced within and among entities (Hamalakis 2017). Once that happens, the function, value, and legitimacy of the assemblage can also change. For example, the typical elements of indigenous Sámi identity include residence in the Sámiland area, traditional livelihoods, family connections, and Sámi language skills. However, nowadays, in the globalized North, only some of the Sámi people make their living through traditional livelihoods; more than 70% of the Sámi live outside the designated Arctic Sámiland, and Sámi language skills vary. As a result, it becomes both personally and politically consequential to ask and decide what configurations are possible or impossible for a Sámi identity assemblage under the current circumstances (Junka-Aikio 2023; Lehtola 2021; Valkonen et al. 2022). This has become an enduring legal debate in Finland with several impacts on the lives and politics of the Sámi people (Kuokkanen 2019; Valkonen 2018).

The focus on interaction between the elements requires a way of thinking that is interested in searching for connections and processes rather than essences. This epistemological stance underlines the shift from essence to interaction, from existence to the process of becoming. The key is not the arrangement or organization per se but the process of arranging elements to create a particular function. This is what the emphasis on the interaction over essence means. Thus, the question is about alliances, not affiliations, becoming rather than being, connections rather than objects. This also means that neither material nor discursive aspects are given primacy but rather their interaction or what they create together (Dovey 2020; Hamalakis 2017). As Buchanan (2015: 390) summarizes: Assemblage "…is the productive intersection of a form of content (actions, bodies, and things) and a form of expression (affects, words and ideas)". This means assemblage, similar to the Foucauldian understanding of discourse, provides one way to overcome the often constructed and unnecessary division between the discursive and the material.

Thus, in examining the Cold Rush, I do not prioritize language or discourse in my analysis; instead, my goal is to examine which elements, be they material or discursive, are assembled to produce a revaluation of Arctic nature resources. In contrast to approaches that focus on a priori chosen components of a complex process, say on language or speakers, the assemblage approach looks at which elements are fitted together in a particular moment and what they produce together. For these reasons, assemblage thinking opens up a new perspective for examining the Cold Rush and the ongoing, incomplete processes of revaluing Arctic natural resources. It highlights how the values linked to Arctic natural resources are never static or fixed. Rather, they evolve over time and under varying circumstances, adapting to economic conditions, ecological imperatives, and political structures. They are multiple and in tension with each other.

From the assemblage perspective, a change in the value of a natural resource requires a change in the whole assemblage regarding the material and discursive elements selected and their relationships. For example, the shift in the value of mining work from a good job to a risky business requires inserting elements of environmental and health risks in the mining assemblage, which were absent previously. Alternatively, we can think of a relatively rapid change in the value of natural gas due to changes in the global political order, especially related to Russia and China. The processual assemblage approach emphasizes the conditions under which a particular value is constructed: the social, political, economic, and historical forces and interests that form an alliance get articulated as a particular kind of value for a natural resource, with particular kinds of relations, privileges, and rights. The processual dimension of the assemblage approach highlights the temporal and spatial situationality of the revaluing process. From this perspective, the revaluing of Arctic natural resources can be examined as interconnected processes that evolve, change, and align with economic conditions, ecological changes, and political structures.

Another productive effect of assemblage thinking for researching a context like Cold Rush is its emphasis on multi-temporality. Historical accounts, often drawing on Foucault (1967/1972), have shown how what appears to be necessary constraints and conditions of our experiences are, in fact, historically contingent. Assemblage takes history seriously but expands beyond the past by including present and possible temporalities in our concepts and practices (Deleuze 1968/1994). The rush for natural resources both draws on and feeds into past, present, and future temporalities. Speculation, however, is the most alluring temporality in the Cold Rush. It is used to make sense of and justify decisions and investments made in the present by referencing a notional—successful and prosperous—future. Speculation combined with boom and bust cycles is a key temporality for people and businesses making a living out of the Rush. This pattern is particularly palpable in the extractive mining industry, where nostalgia for permanent jobs from the industrial era used in arguing for re-opening old (Pietikäinen and Allan 2020).

Also, the agency is thought of as an assemblage, as a redistributed capacity to act across elements in a particular assemblage (Bennett 2010; Roberts 2012). Depending on the function of the assemblage, sometimes it is a human, but not always. Assemblages exist and also function without human agency; the ice turns into water and back again to ice without the direct impact of humans. Similar to new materialist thinking, when thinking about agency in relation to assemblage analysis, the orientation is more towards capacities, potentialities, and affects than focusing on the agency of a sovereign human subject (Braidotti 2019).

This emphasis on process in my assemblage analysis raises two important elements for examining the Cold Rush. First, it underlines that revaluing natural resources has different temporalities. Temporality and resource valuing are inextricable; there is always a temporal dimension to how we understand, value, and experience natural resources. Some of these values can prove to be enduring and resistant to change, while others may be short-lived and ephemeral. Second, the processual approach emphasizes the conditions under which a particular revaluing happens and operates. This raises questions such as: What are the social, political, ecological, economic, and historical forces and interests that form an alliance that gets articulated as a value for natural resource, with particular relations, privileges, and rights?

The way an assemblage is arranged is political not accidental. As Deleuze and Guattari (1987) repeatedly emphasize, the politics of assemblage impact on which elements are selected and how they are organized. Assemblage is not a free-floating, apolitical concept, as some of the critiques suggest (Wallin 2010). Quite the opposite. The concept, like discourse, directs our attention to examining alliances between particular assemblages with power bases and the circumstances under which assemblages become possible, desirable, or marginal (see also Escobar 2008). The power of assemblage lies in its capacity to show us how institutions, organizations, bodies, practices, discourses, and habits make and unmake each other (Wise 2005). Assemblage provides us with an alternative way to examine power dynamics by moving beyond fixed binaries, such as powerful-powerless, central-marginal, majority-minority, which often constrain critical research (see, e.g., Pietikäinen 2016). Thus, it renders our thinking more nuanced and attentive to complex, often messy local conditions and avoids reducing them into a single factor, reason, or solution, or declaring solutions fitting to everyone everywhere (see, e.g., Buchanan 2015).

2.4 The Politics of Arctic Assemblages: De/Reterritorializing Nature

Fig. 2.4 Reindeers by a mine

Assemblages are functional: they emerge when a function or a purpose emerges. Objects, discourses, expressions, and affects are selected and organized for a particular purpose and impact. Therefore, Deleuze and Guattari (1980/1987) suggest that rather than asking only about the essence of the assemblage, for example, "What is this assemblage?", we should also ask about its function, in terms of what it can do and how it works. Assemblages consist of elements from milieus—the surroundings and other contexts—and combined in a particular way for a particular impact (Wise 2005: 78). If we want to understand how an assemblage works, the answer can be provided step by step. By answering questions like: Who are the allies and enemies of this assemblage? What are the consequences and implications of this assemblage here and now? How about there and then? What can the assemblage accomplish, and where are its limits in a particular instance? (Nail 2017: 26). These questions also point to the potentiality of assemblage: by making multiple, perhaps unexpected or so far invisible connections, a productive assemblage creates new means of expression, a new territorial or spatial organization, a new behaviour, or a new realization. Ultimately, assemblage means examining change simultaneously in terms of what is expressed and what is done (Deleuze and Guattari 1987: 504).

The key to understanding how an assemblage functions is that it creates a territory, in other words, it does not function in an empty space. Similar to discussions about space as constructed (Lefebvre 1974/1991). Deleuze and Guattari (1987) argue that territory is not given but constructed, and this construction also entails the organization of the relations between the elements in a particular territory. Their thinking was influenced by the work of a Baltic-German biologist, Jakob von Oexkull (1864–1944), on animals' relationships with their environment. Like animals, assemblages can also be thought of as territorial. Territorial animals, such as Arctic foxes and Arctic grouse, function productively in their own territory but not so well outside their natural inhabitant (Fig. 2.4). Similarly, assemblage only gains its particular function through a particular territory. Without it, assemblage would render as any other complex or messy, random constellation, hard to describe and use analytically. Importantly, territories are more than just spaces: they have a stake, a claim, and they express relations—my berry patch, my home river (see also Wise 2005: 7). Similarly, for a particular territory to be understood as the Arctic, Sámiland, Lapland, or mining belt, Northern light zone or berry patch, it needs to be territorialized as such through the interaction of material and discursive elements. This potentially creates both tensions and possibilities, as we will see later in relation to mining (Chapter 3), superberries (Chapter 4), and Northern Lights businesses (Chapter 5).

One way to get a better sense of why a particular assemblage gets organized in a particular way is to examine the processes through which an assemblage gets stabilized and destabilized. This is what Deleuze and Guattari call the politics of assemblage. They argue that there are four major logics for organising an assemblage: territorial, state, capital, and nomadic.

Territorial assemblage is organized in such a way that the selected elements are coded based on their perceived natural or proper use. The example that Deleuze and Guattari (1987) give is home. A home is not a pre-existing space or the same as a house. Instead, for a house to become a home, it needs to be territorialized. Territorial assemblages divide the world into coded segments. Each concrete element has a designated place, and every life has a plan related to its place in the world: territorial codes define the natural norms of life (Wise 2005.) In a territorialized natural resource assemblage, elements such as land, berries, water, and air are coded to have a natural or proper use and norms regulating how they should be used, when and by whom. So, there is a fixed code for a fixed function. In the context of Cold Rush, the Arctic wild berry assemblage is organized according to this territorial logic with designed berry patches and legitimate pickers, whereas the superberry assemblage, as we shall see, works more with the logic of capital assemblage (see Chapter 4).

State assemblages are organized with relations aiming at unifying or totalizing all selected elements in the assemblage (Deleuze and Guattari 1980/1987: 223). State assemblage is a case of over-coding, that typically uses binaries, circularity, and linearity (Nail 2017: 38–39). Deleuze and Guattari (1980/1987) give statistics and Chomskyan linguistics as examples of state assemblages. In the Arctic context, an example of over-coding is the signifying the Arctic forests as state forests governed solely by the state forest board while ignoring the historical ownership of the indigenous Sámi to these territories. With its homogenising and totalising practices, state assemblage erases multiplicities and tensions related to the forests. In the Cold Rush, one example of state assemblage is the state owned and supported mining business that codes land, infrastructure, labour, and bodies, and the future of the region as depending solely on the future of the mining industry (see Chapter 3).

In the *capitalist assemblage,* the organising logic centres around exchange value. It transcodes the elements and their relationships in terms of exchange value by removing their qualitative relationships and codes to allow them to circulate beyond existing relations (Deleuze and Guattari 1980/1987: 453–455). The codes used in capitalist assemblage establish an economic equivalence between its elements and treat them all as globally exchangeable quantities (Nail 2017: 39–40). The process of re/de-territorialization is critical for the capitalism logic as it allows new value to be created. From this perspective, capitalism operates through the moments of de/re-territorialization, and capitalist assemblage deterritorializes only to bring about a more powerful reterritorialization (Surin 2005: 27). In Cold Rush, attempts to code natural resources only per their exchange value is one of the key processes across the mining industry, berry business, and Northern Lights entrepreneurship.

Nomadic assemblage refers to combinations that are new, creative, and revolutionary for their time or place. In the historical context of Arctic extraction history, environmental issues were new and revolutionary in the 1970s and 1980s. Nomadic assemblages can change and enter into new combinations without arbitrary limits, norms, or so-called natural or hierarchical uses and meanings. Nomadic assemblage is always intermezzo, in-between (Nail 2017: 40–41). For example, the emergence of the Northern Lights business was nomadic in its time in the sense that it combined previously unassembled elements. As a result, it was considered out-of-box and did not fit easily into the existing business structures and practices (see Chapter 5).

These four organising logics are never pure, and they evolve and transform depending on the function and situation. Assemblages are composed of a mixture of these logics to a varying degree, yet one logic can be either more dominant or totally lacking at a particular moment. Examining how these logics play out in a particular assemblage helps analyse its function and conditions.

To examine the politics of assemblage in more detail, Deleuze and Guattari (1987) introduce the interrelated processes of *territorialization, deterritorialization, and reterritorialization*, as well as *de- and recoding*. The process of territorialization refers to stabilising processes, for example, a particular area becoming predominantly understood as a mining belt. However, territories are not fixed for eternity; instead, they are being made and unmade, or in terms of Deleuze and Guattari, they are deterritorialized and reterritorialized. Furthermore, one geographical location can simultaneously host several overlapping territories and territorializations, like the case of the Arctic.

Deterritorialization involves detaching an object, element, sign, or practice from its current context of signification or usage. Deterritorialization suggests a move from the typical norm and ritual towards new change and creativity. The story of how the Huskies became placed in the Finnish Arctic tourism landscape is a telling example of de/reterritorializing impact of consumer power. Unlike in many other circumpolar regions, dogs in the Finnish Arctic were mainly used for hunting and reindeer shepherding and rarely used as sled dogs. Traditional Sámi dog breeds include Finland's Lapland dog (*Suomen lapinkoira*) Lapland's reindeer shepherd, (*Lapin porokoira*) and Sweden's Lapland dog (*Ruotsinlapinkoira*) (The Saami, a Cultural Encyclopaedia 2005). However, as more international tourists started to arrive in the 1990s, they asked for Husky programmes since these dogs belonged to their imagined Arctic nature experience. To meet market demand, some entrepreneurs imported Husky dogs to Finnish Lapland, and today, taking part in a Husky safari belongs to the tourist bucket list. These dogs were deterritorialized from their original territories and reterritorilized into the Lappish tourism territory. This movement reconfigured their function, too: they became part of tourism products. In addition, the importing of the huskies was part of the reterritorializing of the Arctic as a tourism space for global consumption. As this example shows, deterritorialization tends to be followed by reterritorialization in another domain and practice. Deterritorialization is a separation from a given purpose, and reterritorialization is a re-purposing in another domain. The dynamics of de- and reterritorialization processes underline the potentiality of assemblage thinking for political and economic projects. For example, Arctic, Lapland, Sámiland, and periphery assemblages of the same geographical area make different claims for political organization, rights, and redistribution, in relation to the nation-state or global markets. De/reterritorialization reconfigures power relations.

De/recoding is an important aspect of reterritorialization processes. Deleuze and Guattari (1987) use the concept of code to refer to a process of signification that employs not only linguistic codes but also signs and symbols. Like discourse, codes enable and limit what can be expressed and impose order and structure on meaning-making. However, an important extension for the use of codes is their close relation to the processes of de/reterritorialization. In the process of decoding, the code breaks free from the constraints of an existing code, while the process of recoding involves the creation of new codes or the establishment of new territories of meaning. The concept of code underscores the dynamic and evolving nature of signification and language, highlighting the potential for creative and innovative expressions when codes are decoded and recoded (Smith 2011).

Importantly, codes capture not only meaning but also affective intensities. An example of this is the recoding of reindeer. Traditionally, reindeer belong to the traditional livelihoods and cultural practices of indigenous Sámi people. In the tourist imagination, reindeer are coded with the Arctic and the Sámi people, but also with the Disney version of Christmas and Santa Claus, who travels with flying reindeers, particularly with Rudolf, the Red-nosed Reindeer, "the most famous reindeer of all" according to the famous song dating back in 1949. More recently, the hugely popular Disney movie Frozen (I and II), located in the Arctic with indexes to Norway and the Sámi people, depicts a reindeer with the Norwegian name of Sven, described as a loyal friend with a heart of gold. These recodings of reindeer, stemming from popular culture, lend human-type characteristics and experiences to reindeer as well as increasing affective intensities around them to an extent that many tourists, especially children, even when their diet includes meat, refuse to eat reindeer meat, which is offered as a local, sustainable and traditional food.

As highlighted previously, assemblages do not always function as planned and despite collecting elements that are expected to produce the desired outcome, the productive interaction never happens. One example is how Santa Claus never gets his English accent right for the international visitors queuing to him in person at one of the Arctic tourism resorts in Finland. The Finnish Arctic is a booming Christmas tourism destination, especially for British tourists who come for a few nights to see Arctic wonders, including a chance to talk with Santa Claus in person. This brief interaction follows a well-established, scripted interaction frame where Santa Claus asks if visitors have been good and dots in a few details about the visitors or their Christmas wishes, information that Santa's little helpers have gained from the visitors when entering the Santa's office and deliver to Santa's earpiece. Despite collecting all the iconic elements of Christmas together—the winter wonderland, Santa's little helpers, Christmas Carols and decorations, Santa Stories, and Christmas presents and embodied Santa visually coded to resemble an image of Coca-Cola Santa in a red costume with a big white beard and fluent English skills—many visitors leave with some doubt as to whether they have only met a fake Santa.

One of the reasons for this is possibly Santa's accent, which often appears to be a deal breaker, disturbing the Christmas fantasy of meeting Santa in person. Santa's accent does not align with the existing coding of how Santa sounds to each visitor, who may have a different perception of Santa's accent based on movies, childhood stories, and experiences. Santa speaking English with a Finnish accent is difficult to reterritorialize in the terrain of Christmas-related personal affects and memories rooted in various geographical and cultural locations. The service providers have tried to fix this by employing people who speak English as their first language to impersonate Santa, but again, the particular accent that these imported Santa's happened to use did not align with the perceptions of Santa's sound for all visitors. Santa's accents are doomed to sound almost always wrong in the visitors' ears, making successful reterritorialization of Santa impossible. Service providers have added new elements to their Santa assemblage to manage this friction while continuing to offer this tourism product. In the revised Santa assemblage, Santa becomes a multilingual figure with the capacity to speak a few ritual phrases in the visitors' languages. Also, the narrative about the magical nature of Christmas and Santa is emphasized (anything can happen, nobody really knows how Santa works), inviting the visitors to keep the unspoken contract of pretending to believe in Santa.

Importantly for examining the Cold Rush, the re/de-territorialization processes connect assemblage to the conditions and critique of capitalism. Deleuze and Guattari (1980/1987) argue that deterritorialization is a key force of capitalism. It disrupts established meanings, uproots individuals, and re-inscribes work in terms of its exchange value. Perhaps the ultimate reterritorialization of capitalism is that it assigns monetary value to everything. Capitalism deterritorilizes and decodes in order to achieve a more powerful reterritorialization based on capitalist logic (Surin 2005; see also Munro and Thanem 2018). To examine and critique the fixed structures and oppressive forces of capitalism, Deleuze and Guattari (1980/1987) developed a socio-historical analysis of the control of capitalist society called schizoanalysis. In the context of the Cold Rush and revaluing natural resources, capitalism depends on well-functioning processes of de- and reterritorialization of objects, discourses, and desires. At the same time, these very same processes can be used to break from the capitalist profit codes, presenting a possibility for a change and new values for natural resources.

2.5 Cold Calculus: Discourses of Arctic Natural Resource Governance

Fig. 2.5 Arctic frontier

Parallel claims such as "Indigenous people never owned land" or "both mining industry and nature protection can co-exist in harmony" are often used in public debates regarding governance of Arctic natural resources. They are examples of historically enduring discourses that offer one rationale and justification for a particular kind of Arctic natural resources governance. They are also examples of the intertwined relationships between discourse, knowledge, and power (Fig. 2.5).

Discourse, in Foucault's terms (1966/1970), is a key concept for understanding how meanings, power, and knowledge are intertwined. Foucault (1966/1970: 165) describes discourse as the "tangled web of statements" that serve specific social functions, what Foucault called their "enunciative function": they define, delimit, and construct the objects about which they speak; they signal the subject positions from which they are posited; they indicate the broader co-existent network of statements in which they operate; and, finally, they index the institutional settings by which they are sustained (Foucault 1967/1972: 49). Further, discourse can emerge and endure as a body of knowledge comprised of material truth claims rather than as a mere collection of signs (see Morar et al. 2016). This means that, like assemblage, discourse is also both material and symbolic, and can have an impact on the present.

Discourses are more than ways of thinking and producing meaning (Foucault 1967/1972: 48–49; Weedon 1995: 108). It is this "more" dimension that is critical for understanding the revaluing of Arctic natural resources. Drawing on Foucault's work on discourse, power, and knowledge, I understand discourses as both symbolic and material, historically produced, politically invested, and interactionally enacted. Importantly, Foucault argues that historical discontinuity is not simply a matter of a change in the content of knowledge nor a change in what is said. Rather, the change takes place on the plane of the material and discursive conditions for the formation of the discourse (Foucault 1967/1972; Morar et al. 2016). This resonates with assemblage thinking: both Foucault and Deleuze and Guattari insist that we need to pay attention to the specific and situated material, social, and political conditions and rationalities that govern what counts as desirable knowledge or legitimated truth in a given moment and place. Both the material and the discursive matters.

As soon as Foucault published his work in the politically charged France of the 1960s and 1970s, his argument that discourse constitutes knowledge radically questioned the timeless, universal character and position of scientific knowledge (Oksala 2008: 48). Foucault's point is clear from the start of one of his key publications, *The Order of Things* (1966/1970) where he cites an alleged ancient Chinese scheme of classification, which Jorge Luis Borges (1899–1986) had used in his essay *The analytical language of John Wilkins* (1937/1952):

> Animals are divided into a) belonging to the Emperor, b) embalmed, c) tame, d) sucking pigs, e) sirens, f) fabulous, g) stray dogs, h) included in the present classification, i) frenzied, j) innumerable, k) drawn with a very fine camelhair brush, l) et cetera m) having just broken the water pitcher, n) that from a long way off look like flies.

The epistemological implications of this categorization of animals were evident to Foucault. Firstly, it illustrates the power of words: how words create order and hierarchy. Words limit, break, unite, enable, and mobilize. They bring people and elements together and divide them; they include and exclude. This power of language has been extensively studied in fields such as critical linguistics, critical discourse studies, and critical sociolinguistics, as well as in anthropology, gender studies, politics, and philosophy, where the ways in which words matter in terms of, e.g., identities, rights, recognition, and redistribution. The power of language explains why it matters whether a person is labelled a local or outsider, a region is referred to as a periphery or an economic centre, why derogatory terms can be a violent act, or in many judicial systems, a criminal act. The power of language explains why companies, institutions, political movements, and lobbyists invest so much time and money in developing their brands and slogans or how language plays a crucial role in greenwashing resource extraction. It also helps explain how various political parties and movements strategize with language to resignify what human rights or climate change mean, or, alternatively, to reduce their responsibility for political violence by language-washing their actions and dismissing them as "just language" or naming the critique they received as "political correctness". All of this speaks volumes about the power of language. Foucault's work reminds us how speaking and writing are social acts which are embedded in historical and material conditions and which have real, tangible, material, and affective consequences.

From the Foucauldian perspective, discourse is more than language as it shows the intertwined relationship between knowledge, power, subject formation, and truth. Discourses are language plus, and in that sense, examining discourses about revaluing Arctic natural resources rather than the language of revaluing Arctic natural resources aligns with my aim to examine the interaction between material and discursive aspects of revaluing natural resources embedded in processes and structures of power. It does not make sense to me to pull out the language thread from the fabric of the social world and focus only or primarily on the linguistic aspect of social issues. On the contrary: with the help of assemblage, I will try to see language embedded in the multiple other threads of social life and examine their interaction. While discourses have linguistic or other semiotic articulations, their power does not come from linguistic structures or word choice alone but from their alliance with power relations and knowledge production. Discourses can be imagined as having a double helix organization: they have both linguistic and social dimensions, but it is the interaction between these two that makes discourse a dynamic concept that bypasses the traditional micro/macro or symbolic/material split (Pietikäinen and Mäntynen 2019).

And then there is more. Another epistemological take away from the old Chinese categorization of animals is that it shows how no one era, philosophy, or scientific taxonomy is less arbitrary or truer. Rather, concepts and categories from any time or place enable and limit the extent of what can be said and thought. Foucault (1967/1972) uses epistemes or discursive regimes to refer to the rules and norms regulating what is socially possible to say about a specific topic in a given place and time. Each regime imposes its own conceptual constraints and possibilities. It is in these regimes of truths that people, resources, territories, and objects become categorized, hierarchized, legitimized, and valued. Through his study of the history of madness, Foucault shows how what has become seen and understood as "mad" has changed over time and place. And, how, once the classification and definition of madness change, so too do the practices and power relations around them. What to do with a "mad" person depends on how madness is understood: if it is understood as sickness, crime, deviance, divine, the mad person needs respectively to be cured, locked up, punished, medicated, or celebrated. Similarly, the Arctic has been understood, for example, as the end of the world, the last frontier, a periphery, an economic resource, and indigenous land, and consequently, it can be concurred, commodified, colonized, and reclaimed. The different categories for the people living in the area—Lapps, Finns, Sámi, Laplander, Southerners, locals, outsiders, newcomers—each with different perceptions of their rights, tell the story of the intertwined relationship between discourse, power, and knowledge.

This is the power of discourse: it impacts how we are able to see the world around us and to speak about it. It is this combination of seeing and speaking that makes discourse so powerful. From this perspective, knowledge, including scientific knowledge, does not work outside of economic, political, and personal interests and investments, nor outside of historical and material conditions (see, e.g., Cusicanqui 2012; Heller and Duchêne 2011; Heller and McElhinny 2017; De Melo Resende 2021; Lorente 2019; Rambukwella 2019). This perspective includes the Foucauldian idea that discourses have material power and systematically form, shape, and change the definitions of objects. The ways in which knowledge and power are entangled in historically specific circumstances form complex dynamics of what Foucault termed power-knowledge. These dynamics not only impact what is considered as truth/knowledge, but they also lead to subjectivation, the historically specific classification, and shaping of individual human beings into various types of subjects (see, e.g., Angermuller 2023; Lazar 2023; Milani and Richardson 2023).

This view of discourse has several important implications for understanding the revaluing of Arctic natural resources as complex, ongoing, and emergent processes embedded in power dynamics. First, the Foucauldian understanding of discourse emphasizes the multiplicity of discourses. There is no single story or history here but several discourses, each claiming to be the "fact" or the "truth" of the matter. Thus, there is no a priori, "real" or "authentic" value for natural resources, but rather, value becomes the object of empirical research: How do the resources become valued, under which conditions, and by whom? What kind of discursive and material work does this revaluing need? Also, the relationships between various competing discourses become important: Which of the discourses gains more visibility and becomes a norm and a basis for decision-making? And which discourses are left on the margins, subsumed, or even silenced? One way to examine this further is to apply the concept of order of discourse (Foucault 1967/1972) or to understand discourses as forming an emergent rhizome (see above).

A Foucauldian view on discourse challenges the transcendental status of knowledge (Bacchi and Bonham 2014) and renders unspoken governing rules visible. Foucault showed how knowledge is formed by interacting plural and contingent practices within different sites, each involving the material and the symbolic. This argument echoes Deleuze and Guattari's claim that assemblage is an emergent interaction between material and discursive elements. Foucault uses the terms episteme or discourse practices to refer to a historically and culturally specific set of rules for organizing and producing different forms of knowledge. Discourse practices, as developed by Foucault, refer to these knowledge-formation operations of discourses, and they shed light on how specific knowledges (i.e. discourses) operate and the work they do. This approach bridges a symbolic-material distinction and signals the political nature of what is understood as real or the truth in any given time or place (Bacchi and Bonham 2014: 174.)

Understanding discourses as multiple and interconnected with situated rationalities and "truths" leads to the second important implication, namely that discourses, and more broadly the discourse/knowledge/power nexus, are processual and emergent (Rojo 2016). This connects to the key argument in Deleuze's and Guattari's work on assemblage. Thus, discourses are not enduring, unchanging properties or qualities of an event, institution, or time but historically evolving and situationally enacted. Therefore, examining discourses involves following the traces they leave across temporalities and spaces, the routes they travel, and how they get intertwined with social action, political events, and actors. One way to do this is through historiographical methods (Oksala 2008: 48) following from the focus on "the history of the present" in Foucault's archaeological work. Another way is to complement this with ethnography (Heller et al. 2018), especially when investigating conditions and consequences of the intertwined relationship between knowledge construction and power relationships to communities and individuals. In examining Cold Rush, I draw on Foucault's work to examine the role of discourses in creating and changing particular regimes and rationalities related to revaluing Arctic natural resources.

The multitudinous, processual, and emergent nature of discourse in Foucault's work aligns against conformity or a single story, history, or truth. As Oksala (2008: 5) summarizes, for Foucault, research is "a critical practice that relentlessly questioned dogmatic beliefs and intolerable practices in contemporary society". Locating my research on the Cold Rush within this critical context, I see the past, current, and emerging values of natural resources as constructed and having trajectories and histories. Thus, they are themselves objects of change, however gradual, and sometimes a target of explicit political innovation, intervention, and struggle. Just as knowledge and concepts can be constructed, they can also be dismantled, reorganized, and resignified. In exploring this multiplicity and the connectivity of natural recourses in more detail, the work by Deleuze and Guattari, both contemporaries of Foucault, is helpful, especially the concepts of rhizome and assemblage.

2.6 Unfolding Cold Rush: Applying Critical Assemblage Analysis

Fig. 2.6 Unfolding the Cold Rush

> This is how it should be done: Lodge yourself on a stratum, experiment with the opportunities it offers, find an advantageous place on it, find potential movements of deterritorialization, possible lines of flight, experience them, produce flow conjunctions here and there, try out continuums of intensities segment by segment, have a small plot of land at all times. (Deleuze and Guattari 1980/1987: 161)

Reading almost like a poem, this guideline by Deleuze and Guattari (1980/1987) captures the spirit of how to explore the social world with the help of their work. To me, it reads simultaneously as both very practical and very abstract: the rhythm of the process is clear, as are the actions with which to engage, whereas finding the practical details remains open. At the same time, the text opens up a horizon of possibilities, new openings, and becomings. It creates a new headspace to think, write, and explore the Cold Rush (Fig. 2.6).

Works by both Foucault and, Deleuze, and Guattari are often considered as "high" philosophical theory without very clear or analytically specific methodological contributions (e.g., Coleman and Ringrose 2013: 1). Yet both have had an enormous impact on research practices, starting with ontological and epistemological implications and commitments. In my research, Foucault's work has been imperative for conceptualising and analysing the power of discourse, on the one hand, and the historical embeddedness of discourses, on the other. A Foucauldian perspective helps map, trace, and examine how some regimes of truth get valued as normal, preferable, desirable, or dangerous, and what kind of rationality encompasses such discourses. The work by Deleuze and Guattari is crucial for shifting a focus from representation to production by examining complex, ongoing, and emergent processes and their connections with the help of concepts such as assemblage and rhizome. Both concepts try to capture the multiplicity and processual aspects of the social issues we examine; the interaction between their material and discursive dimensions, the dynamics between stability and change, and their connections and temporalities. Assemblage tries to avoid reducing such multiplicities into one dimension. Instead, the focus is on the interaction of these various elements and what they produce together.

The scope of research expands beyond one element in a specific context to investigate the interaction of various elements across moments to understand how they function and what they produce. Language, for example, cannot be isolated or carved out from the material and discursive elements of the assemblage nor form the conditions that give rise to or limit a specific assemblage. We also need to map, trace, connect, and explain the different forces and elements that together produce the object under study, such as the revaluing of Arctic natural resources (Heller et al. 2018). Instead of zooming in on one dot, the assemblage approach invites us to look beyond and between the dots to their connections and processes that they are part of and what they produce together.

Assemblage thinking responds to the need for conceptual and methodological tools that are apt for examining unfolding, complex, intertwined, and ongoing processes and constellations. In practice, assemblage analysis requires working across sections, perspectives, and positions of knowledge production, finding unexpected allies, or employing concepts that, from the start, see our research objects and knowledge production as heterogeneous products of joint forces. Assemblage is a powerful critique of universalist essences and fixed categories, relations, and directions of movement (McFarlane and Anderson 2011). Assemblage thinking asks how entities are able to sustain their connections or "working arrangements" even as internal forces continually threaten to disrupt or destroy equilibrium (Thompson et al. 2022). The assemblage approach also underlines the critical, political aspect of connections and relations by insisting that desire is a driving force for various alliances and arrangements.

Assemblage has great potential to push forward research on complex interactions between language and society. It can override the boundaries of any particular code, category, or perspective, which may colonialize our systems of thinking or focus of analysis. Assemblage challenges us to engage with multiple perspectives, emergent practices, and speculative futures. In my experience, assemblage both helps and forces thinking beyond language to take temporality as an important context and to engage with multiple perspectives and knowledges. In researching the Cold Rush, assemblage has been imperative for keeping an ontological eye on multiplicity and complexity. It has helped me shift the focus from essences and binaries to connections and relationships while accounting for the politics of organization of various elements relevant to the transformations that I have examined. At the same time, assemblage is a challenging concept that opens up so many directions that it becomes challenging to imagine the practicalities of a research project. An antidote is to keep the assemblage situated. The power of assemblage is intertwined with specific political economies, local social histories, and discursive practices. In my research, this translates into combining assemblage thinking with Foucauldian discourse analysis and ethnography.

With these perspectives, assemblage becomes simultaneously grounded and flowing.

Together, the concepts of assemblage, rhizome, and discourse offer possibilities to move beyond the traditional tension between order and structure, endurance and change, essence and process, material and discursive. They help in encompassing and examining the multiplicity and the unfinishedness of the Cold Rush. As highlighted in the previous chapter, in this book, I try to bring these concepts into a dialogue and test a framework that I call *critical assemblage analysis*. I am sure this attempt is far from complete, but I hope it helps in moving beyond fixed positions and ways of working and to help in developing the next steps in examining complex, ongoing, yet critical, and highly political processes in the social world.

The key ideas informing critical assemblage analysis discussed above can be summarized in the following five connected junctures in the research design. Following the rhizomatic ontology, the goal is not to conform; instead, the idea is to become unsettled and probe possibilities. The key ideas and steps for designing critical assemblage research include:

1. *Ontological eye on multiplicity.* Assemblage thinking rejects the idea of an overarching, universalist conceptualization of events and processes, their spaces, and temporalities. Rather, it starts with the multiplicity of social life and its complex potential for various possibilities.
2. *Focus on processes.* Following from this starting point, instead of focusing on one event, action or dimension, an assemblage approach maps ongoing, enduring, and emergent processes related to an assemblage. This focus foregrounds the dynamics and tensions between stability and change.
3. *From binaries and representations to connections, entangelments and production.* The next move is to go beyond binaries and dichotomies and to look for and map connections and relationships between elements of an assemblage and what they produce together. This move is vital as assemblage is empowered by the interaction of its elements.
4. *Politics of function and organization.* As both Foucault and Deleuze and Guattari forcefully argue, how discourses and assemblages are

organized is a matter of power dynamics. Therefore, an analysis of the function and the rationale organising an assemblage is a key to a critical analysis of assemblage. This entails also, especially in relation to capitalism, analysis of re/deterritorializing processes to understand how an assemblage functions. The analysis of re/deterritorialization can be complemented by an analysis of de/recoding.

5. *Ethnography of unfolding processes*. The conditions and consequences of assemblages are intertwined with specific political economies, local histories, and cultural practices. This makes ethnography a productive alley to explore the ways in which any given assemblage is understood and experienced by people in their everyday practices. Ethnography resonates with the idea of probing articulations and experiences of a particular assemblage as well as its specific entanglements with large-scale socio-political, historical, and economic processes. Ethnography helps in seeing how an assemblage functions in practice, what it accomplishes and what its limits are in a particular moment.

Methodologically, ethnography resonates with assemblage thinking as it aims to follow the chosen processes as they emerge and unfold in people's everyday practices and experiences. In the case of Cold Rush, I have been examining the processes of revaluing natural resources related to minerals, water, wild berries, and Northern Lights in moments when there has been an attempt to transform them from natural resources into commodities and businesses: into mining jobs, superberry products, and "Aurora chase" tourism products. These revaluing processes are emergent and unknown, which translates in the ethnographic fieldwork into following the processes where they may lead: some led me to obvious places (like berry patches, mining companies, Aurora business premises), some to unknown or unexpected places (like archives of mining associations, a municipality's economic development project, or food service facilities). The result of the fieldwork is mapping a Cold Rush terrain, a rhizome including the hotspots for these transformations and the connections between them.

From an assemblage perspective, each Cold Rush hotspot becomes an assemblage, which can be examined as an interaction between the material (stuff, objects) and the discursive (discourses, ideas, affects) in a particular moment. This means that there is a no a priori decision to focus, let's say, on discourses or multilingualism, but the aim is to investigate which elements are brought together for a particular function and what they produce together in a given situation. In this way, critical assemblage analysis provides a lens to examine heterogeneous elements and explore the multiple interactions that shape the value of Arctic natural resources.

Assemblage has become a widely used concept across disciplines, including human geography, urban studies, design studies, arts, and education (see, e.g., Baker and McGuirk 2017; Buchanan 2021; Kuismin et al. 2024; Thompson et al. 2022). In discourse and language studies, the concepts developed by Deleuze and Guattari have been taken up recently to address, for example, the relationship between symbolic and material, as well as in discussions on the concept of language (see, e.g., Gurney and Demuro 2023, 2022; Heller and McElhinny 2017; Hovens 2023; Pennycook 2017; Pietikäinen 2016, 2021, 2022; Sharma 2023). As the concept has been applied in different contexts for different purposes, it has gained new usages and meanings, to the extent that some critiques note that it seems to be used synonymously for anything complex, processual, or interconnected, and therefore, the concept is not providing anything new or helpful (Buchanan 2021).

To avoid such a general use of assemblage but rather make use of its potentiality, I try to work closely with an original conceptualization of assemblage by Deleuze and Guattari. Thus, I examine Cold Rush as a rhizomatic assemblage with material and discursive dimensions, organized by different politics for a particular function, and with desire as a driving force (Pietikäinen 2021). I will also look at the unfolding of the processes of re/deterritorialization and re/decoding which results in shifts in the revaluing Arctic natural resources. I pay particular attention to moments of transition in revaluing mining, water, berries, and northern lights. Assemblage is a powerful concept that helps discover what kind of arrangement under which conditions produce the Cold Rush and its ramifications.

The critical assemblage analysis is not alone in its attempt to go beyond binaries to connect multiple perspectives and frameworks. Various prefixes ranging from bi-, multi-, trans-, inter-, super-, post-, and de-, together with the shift from static nouns to active processes (e.g., from language to languaging) in language research and elsewhere, show a need and desire to expand research from one to many, and from fixed to movement (e.g., Esposito 2023; Ke-Schutte and Babcock 2023; Pennycook 2016; Pietikäinen et al. 2008; Pietikäinen 2021; Reyes 2014; Rivera Cusicanqui 2019). Also recent critical sociolinguistic and discourse analytical work related on decoloniality and Southern epistemologies provides productive ways to problematize and renew existing ways of thinking the relationship between language and society (see, e.g., Deumert et al. 2021; Deumert and Makoni 2023; Heugh et al. 2022; Lazar 2020; Makoni et al. 2022; Macedo 2022; De Melo Resende 2021).

We also have well-known analytical frameworks providing tools for empirical work examining the relationship between language and society. For example, in discourse studies, the framework of *nexus analysis,* developed by Ron Scollon and Suzanne B. K. Scollon (2004), brings together ethnography, interactive sociolinguistics, and discourse analysis to examine social action. Similar to the idea of assemblage, in nexus analysis, the key idea is that the co-function of different elements—the body, discourses, and the interaction order—makes a social action. Also, the widely used three-dimensional framework by Norman Fairclough (1992) connects text analysis with in-depth contextual analysis of discourse practices and social conditions to examine multi-dimensional discourse. Similarly, multi-sited ethnography connects everyday practices and experiences with historical, political, economic, and cultural processes (Heller et al. 2018).

Critical assemblage analysis produces alternative ways to explore the dynamics between language and society. It moves beyond language in an attempt to engage with the multiplicity of social issues we examine their material and discursive dimensions, their relative stability and dynamics, their different relationships and temporalities, and their embeddedness in power relations. Seeing language co-existing and co-functioning with the other elements in an assemblage opens up new possibilities, alliances, and solutions for research. Furthermore, critical assemblage analysis brings about a crucial ontological shift from essences, binaries, and layers to the more complex terrain of multiplicity, heterogeneity, convergence, and flows. While dualism and binaries—the *tree structure* in Deleuzian terms—may continue to be useful in some contexts, they are limiting in others. Critical assemblage analysis responds to the need for conceptual and methodological tools that are appropriate for examining real-life, complex, messy, intertwined, and ongoing processes and constellations, *rhizomes* in Deleuzian terms.

Methodologically, assemblage thinking also suggests a move from representation to production (Adkins 2015; Henderson et al. 2016; Smith 2011). Adkins (2015: 29) illustrates the difference between representation and production by comparing stickers and drawings: ready-made stickers reproduce the same image over and over again, even when placed in new contexts, while drawings have critical potential to produce new images and connections. The power of drawing resonates with the recommendation by Deleuze and Guattari (1980/1987:11–13) to make a map oriented towards experimentation, as a means of finding new terrains and fostering novel connections. Methodologically, the idea of creating a map of a rhizomatic Cold Rush assemblage aligns with the goal of assemblage thinking, which is to focus on processes of discovery.

Aligning with the logic of qualitative research, I do not think that it is possible to provide detailed recipes for how exactly to do a productive analysis across specific times and places. Qualitative research, at its best, provides a detailed account of a specific situation but also makes connections to wider processes, phenomena, and structures, thus explaining what happens in a particular moment and how it matters (e.g., Heller et al. 2018). In my view, qualitative analysis renders an examination of a rhizomatic assemblage that produces a particular function under particular conditions. However, what can be done is to share the probes and steps of the qualitative research process and the thinking tools used in the process. These steps and tools may or may not work in your case, but at least they may push your thinking into new terrains.

Below, I summarize the critical assemblage analysis in the way I have used it in researching the Cold Rush with the help of key questions and a drawing that maps out the key elements of the analysis and their relationships (Fig. 2.7).

Fig. 2.7 Critical assemblage analysis, drawing by Ilmari Hakala

(1) **Elements**: What elements are selected for a particular assemblage?
(2) **Politics**: How is the assemblage arranged? What is the rationale guiding the specific arrangement? What gets re/deterritorialized? What gets de/recoded? What gets connected?
(3) **Effect**: What does the interaction between the elements produce together? What is the function of this assemblage?

(4) **Moment:** What are the particular conditions and consequences of the assemblage?

Through their work, Deleuze and Guattari develop concepts and thinking that emphasize how things connect rather than how they are. They prefer to consider things not as essences but as multiplicities, focusing on things in terms of unfolding forces rather than static essences. The critical analytical question is always: given the specific situations, what kind of assemblage would be required to produce the desired function.

With their work in mind, in this book, I ask: given the specific situation in the Finnish Arctic, including acceleration of climate change, the reconfiguration of political alliances, and an intensifying economic boom, what kind of assemblages produce Cold Rush, especially related to the revaluing of land, water, and air? And, what are the conditions and consequences of Cold Rush assemblages? In the following chapters, I will try to answer these questions by examining revaluing Arctic mining, Arctic wild berries, and Northern lights. May (2005: 38) uses the metaphor of origami to describe working with assemblage: the paper is never cut but involves a repetitious rhythm of folding, unfolding, and refolding. In the next chapters, I write about the unfolding Cold Rush across mining, berry, and tourism terrains in the Finnish Arctic.

References

Adkins, Brent. 2015. *Deleuze and Guattari's A Thousand Plateaus: A Critical Introduction and Guide*. Edinburgh: Edinburgh University Press.

Angermuller, Johannes. 2023. 'The Analysis of Discursive Subjects'. In *Handbook of Political Discourse*, edited by Piotr Cap, 180–203. Cheltenham: Edward Elgar Publishing. https://doi.org/10.4337/9781800373570.00021.

Appadurai, Arjun. 1996. *Modernity at Large: Cultural Dimensions of Globalization*. Minnesota: University of Minnesota Press.

Bacchi, Carol, and Jennifer Bonham. 2014. 'Reclaiming Discursive Practices as an Analytic Focus: Political Implications'. *Foucault Studies*, April, 179–92. https://doi.org/10.22439/fs.v0i17.4298.

Baker, Tom, and Pauline M. McGuirk. 2017. 'Assemblage Thinking as Methodology: Commitments and Practices for Critical Policy Research'. *Territory, Politics, Governance* 5 (4): 425–42.

Bennett, Jane. 2010. *Vibrant Matter: A Political Ecology of Things*. Durham: Duke University Press. https://doi.org/10.1515/9780822391623.

Braidotti, Rosi. 2019. 'A Theoretical Framework for the Critical Posthumanities'. *Theory, Culture & Society* 36 (6): 31–61. https://doi.org/10.1177/0263276418771486.

Buchanan, Ian. 2015. 'Assemblage Theory and Its Discontents'. *Deleuze Studies* 9 (3): 382–92. https://doi.org/10.3366/dls.2015.0193.

———. 2021. *Assemblage Theory and Method*. London: Bloomsbury Publishing Plc. https://doi.org/10.5040/9781350015579.

Coleman, Rebecca, and Jessica Ringrose, eds. 2013. *Deleuze and Research Methodologies*. Deleuze Connections. Edinburgh: Edinburgh University Press.

Cusicanqui, Silvia Rivera. 2012. 'Ch'ixinakax Utxiwa: A Reflection on the Practices and Discourses of Decolonization'. *South Atlantic Quarterly* 111 (1): 95–109. https://doi.org/10.1215/00382876-1472612.

De Melo Resende, Viviane. 2021. 'Decolonizing Critical Discourse Studies: For a Latin American Perspective'. *Critical Discourse Studies* 18 (1): 26–42. https://doi.org/10.1080/17405904.2018.1490654.

Deleuze, Gilles. 1968/1994. *Difference and Repetition*. Translated by Paul Patton. New York: Columbia University Press.

Deleuze, Gilles, and Felix Guattari. 1980/1987. *A Thousand Plateaus: Capitalism and Schizophenia*. Translated by Brian Massuni. London: Bloomsbury Academic.

Deumert, Ana, Anne Storch, and Nick Shepherd, eds. 2021. *Colonial and Decolonial Linguistics: Knowledges and Epistemes*. Oxford, New York: Oxford University Press.

Deumert, Ana, and Sinfree Makoni, eds. 2023. *From Southern Theory to Decolonizing Sociolinguistics: Voices, Questions and Alternatives*. Bristol: Multilingual Matters.

Dovey, Kim. 2020. 'Place as Assemblage'. In *The Routledge Handbook of Place*, edited by Tim Edensor, Ares Kalandides, and Uma Kothari, 21–31. London: Routledge.

Escobar, Arturo. 2008. *Territories of Difference: Place, Movements, Life, Redes*. Durhamn: Duke University Press. https://doi.org/10.1215/9780822389439.

Esposito, Eleonora. 2023. 'Discourse, Intersectionality, Critique: Theory, Methods and Practice'. *Critical Discourse Studies*, July, 1–17. https://doi.org/10.1080/17405904.2023.2230602.

Fairclough, Norman. 1992. *Discourse & Social Change*. London: Polity Press.

Feely, Michael. 2020. 'Assemblage Analysis: An Experimental New-Materialist Method for Analysing Narrative Data'. *Qualitative Research* 20 (2): 174–93. https://doi.org/10.1177/1468794119830641.

Foucault, Michel. 1966/1970. *The Order of Things: An Archeology of the Human Sciences*. Translated by A.M. Sheridan Smith. London: Routledge.

———. 1967/1972. Archaeology of Knowledge. Translated by A.M. Sheridan Smith. New York: Pantheon Books.

Gehman, Joel, Garima Sharma, and Alim Beveridge. 2022. 'Theorizing Institutional Entrepreneuring: Arborescent and Rhizomatic Assembling'. *Organization Studies* 43 (2): 289–310. https://doi.org/10.1177/01708406211044893.

Gurney, Laura, and Eugenia Demuro. 2022. 'Tracing New Ground, from Language to Languaging, and from Languaging to Assemblages: Rethinking Languaging Through the Multilingual and Ontological Turns'. *International Journal of Multilingualism* 19 (3): 305–24. https://doi.org/10.1080/14790718.2019.1689982.

———. 2023. 'Simultaneous Multiplicity: New Materialist Ontologies and the Apprehension of Language as Assemblage and Phenomenon'. *Critical Inquiry in Language Studies* 20 (2): 127–49. https://doi.org/10.1080/15427587.2022.2102011.

Hamalakis, Yannis. 2017. 'Sensorial Assemblages: Affect, Memory and Temporality in Assemblage Thinking'. *Cambridge Archaeological Journal* 27 (1): 169–82.

Heller, Monica, and Alexandre Duchêne. 2011. 'Pride and Profit: Changing Discourses of Language, Capital and Nation-State'. In *Language in Late Capitalism*, edited by Alexandre Duchêne and Monica Heller, 1–21. London: Routledge.

Heller, Monica, and Bonnie S. McElhinny. 2017. *Language, Capitalism, Colonialism: Toward a Critical History*. Toronto: University of Toronto Press.

Heller, Monica, Sari Pietikäinen, and Joan Pujolar. 2018. *Critical Sociolinguistic Research Methods: Studying Language Issues That Matter*. New York: Routledge.

Henderson, Linda, Eileen Honan, and Sarah Loch. 2016. 'The Production of the Academicwritingmachine'. *Reconceptualizing Educational Research Methodology* 7 (2): 4–18. https://doi.org/10.7577/rerm.1838.

Heugh, Kathleen, Christopher Stroud, Kerry Taylor-Leech, and Peter I. De Costa, eds. 2022. *A Sociolinguistics of the South*. New York: Routledge.

Hovens, Daan. 2023. 'Breakdowns and Assemblages: Including Machine-actants in Sociolinguistic Ethnographies of Blue-collar Work Environments'. *Journal of Sociolinguistics* 27 (1): 3–23. https://doi.org/10.1111/josl.12565.

Junka-Aikio, Laura. 2023. 'Whose Settler Colonial State? Arctic Railway, State Transformation and Settler Self-Indigenization in Northern Finland'. *Post-colonial Studies* 26 (2): 279–301. https://doi.org/10.1080/13688790.2022.2096716.

Ke-Schutte, Jay, and Joshua Babcock. 2023. 'Toward a Non-binary Semiotics of Intersectionality: Linguistic Anthropology in the Wake of Coloniality'. *Journal of Linguistic Anthropology* 33 (2): 112–30. https://doi.org/10.1111/jola.12397.

Kuismin, Ari, Alice Wickström, Joel Hietanen, and Saija Katila. 2024. 'From Dust to Buzz: Reconfiguring Space for Organization-Creation'. *Organization Studies* 45 (4): 523–44. https://doi.org/10.1177/01708406231225041.

Kuokkanen, Rauna. 2019. 'At the Intersection of Arctic Indigenous Governance and Extractive Industries: A Survey of Three Cases'. *The Extractive Industries and Society* 6 (1): 15–21. https://doi.org/10.1016/j.exis.2018.08.011.

Lazar, Michelle M. 2020. 'Politics of the "South": Discourses and Praxis'. *Discourse & Society* 31 (1): 5–18. https://doi.org/10.1177/0957926519886126.

———. 2023. 'Intersectionalisation as Meta-Discursive Practice: Complicated Power Dynamics in Pink Dot's Movement-Building'. *Critical Discourse Studies*, June, 1–18. https://doi.org/10.1080/17405904.2023.2228935.

Lefebvre, Henri. 1974/1991. *The Production of Space*. Translated by Donald Nicholson-Smith. Cambridge: Blackwell Publishing.

Lehtola, Veli-Pekka. 2021. 'Contested Sámi Histories in Finland'. In *Sámi Research in Transition*, edited by Laura Junka-Aikio, Jukka Nyyssönen and Veli-Pekka Lehtola. London: Routledge. https://doi.org/10.4324/9781003090830

Lorente, Beatriz P. 2019. 'Dismantling the Colonial Structure of Knowledge Production'. *Language, Culture and Society* 1 (1): 152–56. https://doi.org/10.1075/lcs.00012.lor.

Macedo, Litiane Barbosa. 2022. 'Intersecting Afroperspective Thinking and Critical Discourse Analysis: Possibilities to Decolonize Discursive Studies'. In *Coloniality in Discourse Studies*: A Radical Critique, edited by Solange Maria de Barros and Viviane Resende. London: Routledge.

Makoni, Sinfree, Anna Kaiper-Marquez, and Lorato Mokwena, eds. 2022. *The Routledge Handbook of Language and the Global South/s*. New York: Routledge.

May, Todd. 2005. *Gilles Deleuze: An Introduction*. New York: Cambridge University Press.

McFarlane, Colin, and Ben Anderson. 2011. 'Thinking with Assemblage'. *Area* 43 (2): 162–64. https://doi.org/10.1111/j.1475-4762.2011.01012.x.

Milani, Tommaso M., and John E. Richardson. 2023. 'Discourses of Collective Remembering: Contestation, Politics, Affect'. *Critical Discourse Studies* 20 (5): 459–76. https://doi.org/10.1080/17405904.2022.2090979.

Morar, Nicolae, Thomas Nail, and Daniel W. Smith. 2016. 'Introduction'. In *Between Deleuze and Foucault*, edited by Nicolae Morar, Thomas Nail and Daniel W. Smith, 1–8. Edinburg: Edinburgh University Press. https://doi.org/10.1515/9781474415095-002.

Munro, Ian, and Torkild Thanem. 2018. 'Deleuze and the Deterritorialization of Strategy'. *Critical Perspectives on Accounting* 53 (2018): 67–78. https://doi.org/10.1016/j.cpa.2017.03.012.

Nail, Thomas. 2017. 'What Is an Assemblage?' *SubStance* 46 (1): 21–37.

Oksala, Johanna. 2008. *How to Read Foucault*. New York: W. W. Norton & Co.

Pennycook, Alastair. 2016. 'Mobile Times, Mobile Terms: The Trans-Super-Poly-Metro Movement'. In Sociolinguistics: Theoretical Debates, edited by Nikolas Coupland, 201–16. Cambridge: Cambridge University Press.

———. 2017. 'Translanguaging and Semiotic Assemblages'. *International Journal of Multilingualism* 14 (3): 269–82. https://doi.org/10.1080/14790718.2017.1315810.

Pietikäinen, Sari. 2016. 'Critical Debates: Discourse, Boundaries and Social Change'. In *Sociolinguistics: Theoretical Debates*, edited by Nikolas Coupland, 263–81. Cambridge: Cambridge University Press. https://doi.org/10.1017/CBO9781107449787.013.

———. 2021. 'Powered by Assemblage: Language for Multiplicity'. *International Journal of the Sociology of Language* 2021 (267–268): 235–40. https://doi.org/10.1515/ijsl-2020-0074.

———. 2022. 'Assemblage of Art, Discourse and Ice Hockey: Designing Knowledge About Work'. *Journal of Sociolinguistics* 26 (5): 553–67. https://doi.org/10.1111/josl.12470.

Pietikäinen, Sari, Riikka Alanen, Hannele Dufva, Paula Kalaja, Sirpa Leppänen, and Anne Pitkänen-Huhta. 2008. 'Languaging in Ultima Thule: Multilingualism in the Life of a Sami Boy'. *International Journal of Multilingualism* 5 (2): 79–99. https://doi.org/10.1080/14790710802152289.

Pietikäinen, Sari, and Kori Allan. 2020. "Jobs for Life?': Mining Temporalities in a Transforming Arctic Periphery'. In *Language, Global Mobilities, Blue-Collar Workers and Blue-Collar Workplaces*, edited by Kellie Gonçalves and Helen Kelly-Holmes, 27–48. London: Routledge. https://doi.org/10.4324/9780429298622-2.

Pietikäinen, Sari, and Christine Hegel. 2021. 'Assembling Success: Desire, Worker Mobility, and Value Creation in Finnish Ice Hockey'. *Anthropology of Work Review* 42 (2): 71–81. https://doi.org/10.1111/awr.12221.

Pietikäinen, Sari, and Anne Mäntynen. 2019. *Uusi kurssi kohti diskurssia.* Tampere: Vastapaino.

Rambukwella, Harshana. 2019. 'On Hybridity, the Politics of Knowledge Production and Critical Language Studies'. *Language, Culture and Society* 1 (1): 126–31. https://doi.org/10.1075/lcs.00008.ram.

Reyes, Angela. 2014. 'Linguistic Anthropology in 2013: Super-New-Big'. *American Anthropologist* 116 (2): 366–78. https://doi.org/10.1111/aman.12109.

Rivera Cusicanqui, Silvia. 2019. 'Ch'ixinakax Utxiwa: A Reflection on the Practices and Discourses of Decolonization'. *Language, Culture and Society* 1 (1): 106–19. https://doi.org/10.1075/lcs.00006.riv.

Roberts, Tom. 2012. 'From "New Materialism" to "Machinic Assemblage": Agency and Affect in IKEA'. *Environment and Planning A: Economy and Space* 44 (10): 2512–29. https://doi.org/10.1068/a44692.

Rojo, Luisa Martín. 2016. 'Language and Power'. In *The Oxford Handbook of Language and Society*, edited by Ofelia Carcía, Nelson Flores and Massimiliano Spotti, 77–103. Oxford: Oxford University Press.

Scollon, Ronald, and Suzanne B. K. Scollon. 2004. *Nexus Analysis: Discourse and the Emerging Internet*. London: Routledge.

Sharma, Bal Krishna. 2023. 'A New Materialist Perspective to Studying L2 Instructional Interactions in Engineering'. *International Journal of Bilingual Education and Bilingualism* 26 (6): 689–707. https://doi.org/10.1080/13670050.2020.1767030.

Smith, Daniel W. 2011. 'Flow, Code and Stock: A Note on Deleuze's Political Philosophy'. *Deleuze Studies* 5 (supplement): 36–55. https://doi.org/10.3366/dls.2011.0036.

Surin, Kenneth. 2005. 'Force'. In *Gilles Deleuze: Key Concepts*. Edited by Charles J. Stivale, 19–30. McGill-Queen's University Press. http://www.jstor.org/stable/j.cttq486p.7.

Thompson, Greg, Sam Sellar, and Ian Buchanan. 2022. '1996: The OECD Policy-Making Assemblage'. *Journal of Education Policy* 37 (5): 685–704. https://doi.org/10.1080/02680939.2021.1912397.

Valkonen, Sanna. 2018. 'Conceptual Governance on Defining Indigeneity—The Sámi Debate in Finland'. In *Knowing from the Indigenous North*, edited by Thomas Hylland Eriksen, Sanna Valkonen and Jarno Valkonen, 142–62. London: Routledge.

Valkonen, Sanna, Áile. Aikio, Saara Alakorva, and Sigga-Marja. Magga. 2022. *The Sámi World*. London: Routledge. https://doi.org/10.4324/9781003025511.

Wallin, Jason J. 2010. *A Deleuzian Approach to Curriculum: Essays on a Pedagogical Life*. Basingstoke: Palgrave Macmillan.

Weedon, Chris. 1995. *Feminist Practice and Poststructuralist Theory*. Cambridge: Blackwell.

Wise, J. Macgregor. 2005. Assemblage. In *Gilles Deleuze: Key Concepts*, ed. Charles J. Stivale, 77–87. London: Routledge.

3

Mining Connections: Mineral Modernity in the Arctic

3.1 Arctic Laboratory of Modernity

Fig. 3.1 Tornio River between Sweden and Finland

S. Pietikäinen, *Cold Rush*, Language and Globalization,
https://doi.org/10.1007/978-3-031-63995-1_3

"We are now at the place where civilisation started in this part of the world", says Pentti, a retired miner with a keen interest in fishing and local history, shouting over the roar of the famous Lappea rapids from the other end of his long, narrow river fishing boat. I look around and see a few houses dotting the banks of the majestic Tornio River and vast forests closing in. I don't see signs of past or present glory, but I know that Pentti is referring to the historical *Kengis* ironworks and the famous *Kengis* manor, which stands just one kilometre up from the rapids. The famous Walloon owners and developers of the Kengis manor and ironworks, two brothers by the name of *Momma*, heralded the arrival of the capitalist system in Arctic mining. For centuries, the Kengis manor was considered as the last outpost for visitors before embarking on expeditions to Arctic frontiers. Pentti likes to surprise fishing tourists with a story about seventeenth- and eighteenth-century explorers sent by the King of France or the Pope in Rome to explore the Arctic frontier. They, too, as the fishing tourists today, travelled along this same river (Fig. 3.1).

The junction of two big Lapland rivers, the *Lappea* rapids, has attracted fish, people, and capital for centuries. The bounty of natural resources—water, fish, minerals, forests—has made two neighbouring villages, *Lappea* (on the Finnish side) and Kengis (on the Swedish side) into the first hotspot of multiple economic and political processes, including colonialization, industrialization, radical religious and communist movements, and environmental projects. The Tornio River, or *Väylä* (the route), as the locals affectionally call it, is a historical Arctic highway that connects these villages into global processes and transformations. The river has witnessed several important shifts in the area's political economy. Place names and archaeological studies show the indigenous Sámi people lived in the area before the Finnish settlers arrived in larger number from seventeenth century onwards (Nordin and Ojala 2017). As settler colonialism expanded, the region became a predominantly Finnish-speaking area, yet was governed by the Swedish crown until 1809. As a result of the so-called Finnish War (1808–1809), Sweden lost Finland and all of its domains east of the Tornio River to Russia, and the Grand Duchy of Finland was established under Russian rule. The Tornio River became a border river, first between Sweden and Russia, and after Finland's independence in 1917, a border between Sweden and

Finland. Yet, the locals continue to identify themselves as inhabitants of Tornio River valley, a community connected by the river. They are also connected by a shared language resource, Meänkieli, traditionally considered a local dialect of Finnish but a recognized minority language in Sweden since 1999 (Mantila 2000; Pietikäinen et al. 2010).

Rivers and mines are intertwined in the valley. The rivers provided water and energy for mining processes, and in the old days, they also provided infrastructure for the circulation of goods and people. The rhizome of a big local iron ore deposit, already discovered in the seventeenth century, spans the river on both sides. So too do the mines. On the Swedish side, there is the historical Kengis Ironworks and the currently operational iron ore mine *Kaunisvaara* (operating since 2018), which has a license to discharge excess water into the *Muonio* River, some 20 km north of the *Lappea* rapids. On the Finnish side, the *Rautuvaara/Hannukainen* iron ore mine has closed, but there are plans for it to reopen and become the largest open pit mine in Europe. In this case, it would discharge its process water into the Muonio River. There are also several active mining development plans across the region on both sides of the river.

At the same time, the Tornio River is the longest free-flowing river in Europe, and it belongs to *Nature 2000*, a network of nature protection areas in the EU. It is an important spawning river for many species of fish, including Baltic salmon. This has made it a popular fishing tourism destination and the object of several EU-funded development projects in the riverside municipalities, including the Finnish *Kolari*. Also, Lappea and Kengis villages are home to several small-scale fishing entrepreneurs, including one on the premises of Kengis Ironworks. This is the same salmon river I spent summers as a "salmon widow", discussed in Chapter 1.

For four centuries, the Kengis Ironworks provided income for the locals, not just mining work but also service jobs, food, and coal production. Many Lappea villagers and others from nearby villages worked in coal production, earning the municipality the name Kolari. The name comes from the Swedish word *kolare*, which means a person who produces coal (kol in Swedish). The most enduring impact of Kengi*s*, however, goes beyond the villagers earning a living. The ironworks, an early modern project and colonial enterprise in the Arctic, connected European colonial expansion to areas perceived as frontiers (Komu 2020; Naum 2018; Nordin 2015; Nordin and Ojala 2017). Kengis ironworks facilitated settler colonialism and paved the way for industrialization and capitalist exploitation of natural resources, which had the ripple effects of creating a class system and political organization of the labour force (Aatsinki 2009).

The ironworks operated within the logic of the emerging capitalist system, introducing new ideas and practices. Many were connected to the strategies of modern governance that used spatial, temporal, and moral orders to shape the locals into docile bodies, complicit under the conditions of industrial work. The ironworks burned down in the late nineteenth century, but the manor continues to be a home to a major Swedish industrial family. The place is still connected to the region's economic development: a few years ago, the mining grounds by the river were turned into a luxury fishing lodge for international fishing tourists. A modern *Kaunisvaara* iron ore mine is located thirty kilometres away from Kengis, operates 24/7, and produces annually 2.2 million tons of iron ore.

The bond between rivers and mines—between nature and extractive industries—creates possibilities and contradictions that I explore in this chapter using the critical assemblage framework. The environmental risks that the mines pose to nature and the rivers, in particular, are the key reasons why the mines in this region are opposed. The value of mining work as an all-year-round job in the Arctic region, which has an aging population and fragile infrastructure, is one of the key reasons the local mines are supported. These opposing views materialize in Kolari in a heated debate about whether or not the old Rautuvaara/Hannukainen mine on the Finnish side should be permitted to be reopened (Hast 2021; Jokinen 2019; Komu 2020). The future of the mining industry in the region divides communities, villages, and families.

The controversial, polarizing aspects of Arctic mines got my attention while listening to people's fishing stories and accounts of the region's economic development. Everybody seemed to have an opinion about the mines and people seemed to be able to hold contradictory values and positions within the same frame: a keen fisher or berry picker could be simultaneously pro-mine and pro-river protection. The gains and risks related to mining were debated over coffee tables, in the municipality meetings, and on the pages of local newspapers. The situation has become so tense that many resorted to silence to avoid an open conflict with neighbours or family members. As Pentti says: "A topic better not to talk about with your neighbours".

These multiplicities and ongoing processes related to the value of mining can be productively examined with the help of critical assemblage analysis. Exploring the slow unfolding of Arctic mining across four centuries, I examine the Arctic mining assembling with help of critical assemblage analysis: which elements are needed and how they are organized for a mine to function productively under changing conditions; how these assemblages re/deterrorialize and recode resources, relations and people; and, ultimately, how they reconfigure relations between people and resources, land in particular.

To understand what is at stake in the mining debates, I mapped the region's 400-year mining history. The historical development of Arctic mining showed how mines were intertwined with the expansion of capitalism and colonialism in the region and later with radical religious and political movements, leading to a cycle of mining boom and bust in the late twentieth century. This creates shifts in assemblage politics: the capitalist assemblage is predictably dominant, but it interacts with the state assemblage. Interestingly, what starts as a nomadic, resistant assemblage, like radical religious and political movements, becomes stabilized and transformed at least partly into state assemblages. In these moments of shift, people, resources, and relations become reconfigured: the Arctic land becomes deterritorialized as indigenous Sámiland and reterritorialized as a state-owned land, people become recoded, e.g., as nomads, settlers, workers, "reds" and "whites".

A red thread across the centuries and mining sites is the trust in the progress that mining is believed to bring about: each new mining investment promises to deliver a better life, if not right now, then at least in the future. Mines are part of the rhizome of local history, current sources of revenue, and future speculations. They also connect the Arctic region to global economic and political processes, which can have a disproportional impact on small communities (see Pietikäinen and Kelly-Holmes 2013), especially indigenous Sámi communities (Kuokkanen 2019).

The mining story starts with the development of the oldest and most influential mining site by the Tornio River, the Kengis ironworks, from the seventeenth century onwards. The colonial processes and capitalist production at Kengis became intertwined with radical religious and political movements in an attempt to reconfigure relations of production and redistribution of profit. This history helps to understand how the Kolari region became and continues to be politically the "reddest" municipality in Finland, with connections to the Civil War (1917–1918), the Second World War (1939–1945), and the industrialization of the region in the 1960s onwards. The nostalgia related to the golden mining era in Kolari from 1960 to 1980 still has resonance for the older generation in the current situation where the environmental risks of mining, expressed mainly by the younger generation, are challenging the traditional value of mining.

Examining the historical and current value of mining in the area has entailed studying historical documents, current strategies, and economic development texts, resulting in a wide range of data. During the 2016–2021, I have taken part in several social license to operate (SLO) events, aiming at securing local acceptance for mining plans, typically organized by the municipalities together with the mining company, and followed the political decision-making process related to mining licensing in the public municipality council meetings and minutes. I have also collected local media coverage about the topic. I have visually documented two mines currently closed in Kolari and their gradual change in the five-year period in anticipation of the potential reopening. To engage with locals' experiences and sense-making related to value of mining, I have interviewed key stakeholders in the region's economic development and retired workers from the local mines. To learn more about what those people who lived through the mining boom and bust in Kolari some forty years back saw the past, current, and future situation of mining in the area, I have had several discussions over the years with three retired mining workers, Pentti, Taimi, and Simo, and this chapter features their experiences. They are all over sixty-five years old, former workers for local mines, and now spending their retirement time by the Tornio River. During the summers, we all live in the same Kolari region and often meet by the Tornio River. Pentti travels with us through the whole historical development of the mines as he also has an avid interest in the local history. We will hear the experiences of Taimi and Pentti towards the end of the chapter when the golden era of the Kolari mining boom in 1960–1980 is discussed.

Pentti, a retired miner, and a keen fisher, is known for his supreme fishing and hunting skills, but also his interest in local history and politics. Besides working in various mines in the region, he has also worked for the Finnish Museum Association as a fieldworker cataloguing historical sites related to mining and old Lapland borders among others. But closest to Pentti's heart is the wild Arctic salmon that still lives in the Tornio River. He lives, breathes, and talks salmon. During the short Arctic summer, he spends all his waking hours "at the river" in his long, graceful river boat with a powerful engine, going up and down the river rapids, casting his meticulously tended rods with carefully selected bait in the water. As the summer progresses, he becomes sun-tanned and muscular, almost languishing. "Thank God there is Monday (fishing is forbidden) so that I can shop, cook, do laundry, and rest", Pentti laughs. Due to his skills and hours spent at the river, he catches more fish than anyone else. In the winter months, he works tirelessly to protect wild salmon for the local fishing rights and the river's wellbeing. He writes petitions, responds to the fishing regulations, and takes part in an endless stream of meetings about fishing in the Tornio River. In fishing circles, Pentti is famous for his talent, stamina, and knowledge and is regularly interviewed for fishing documentaries and media stories. Many want to fish with him, and I know I am privileged to get an invitation to accompany him. My strongest skill in this context is being able to be patient and quiet for as long as the fishing trip lasts. And he enjoys having an audience for his stories, too.

Taimi and Simo, a couple who have spent more than half of the century together, began their shared in an Arctic mine. Mining jobs offered them a new kind of independence and freedom as well as a possibility for a housing to raise their family. Taimi worked in the mine administration, and Simo belonged to the middle management of the mine. They have lived through the boom and bust of the region's mines. Just before the mining bust really hit the region in the late 1980s, they were able to move into new jobs in the region. Now, as pensioners, they live an active life with various cultural and sports events in the region: they go to the rare concerts organized nearby, ski several hundred kilometres every year, and both like to fish by the river. They are also keen berry and mushroom pickers and follow the seasonal changes in nature closely. I regularly visit them at their well-cared for second home by the Tornio River, and we discuss the berry and fish seasons as well as the latest mining developments over a cup of coffee, the most typical social arrangement in the area.

3.2 Colonial Routes and Rhizomes of an Arctic Mine: *Kengis* Ironworks (1646–1879)

Fig. 3.2 The *Kengis* Manor

"I actually have some Walloon heritage", my future husband declared on our first date in a southern part of Finland in the early 1990s. At the time, I thought that he was just trying to make an impression as, back then, coming from Lapland was generally not considered as a cool resource for distinction as it is today. In fact, from my southern perspective, Lapland looked like an emptying, peripheral place that people left for jobs, schools, and an urban lifestyle. But actually, he was narrating part of his local Lappish identity, rooted in a centuries-long international mining history and labour migration. *Lappea,* the Tornio riverside village my husband comes from, is only two kilometres from the Kengis Manor and ironworks, the northernmost ironworks with a 200-year existence (1646–1879) (Fig. 3.2).

The ironworks was founded in 1646 as part of the expansion plan of the Swedish iron industry, the engine of the country's prosperity for centuries. Accelerating colonial trade drove global markets for metal products, and Sweden sought to profit from this. The Swedish metal exports were already considerable in the Middle Ages, but in the seventeenth century, the production of iron, copper, and brass dominated global trade, with Sweden producing metals and alloys for colonial extraction and trade in cotton production, sugar plantations, the fur trade, and the gold trade in the Americas (Nordin 2015: 197; see also Nordin 2024; Sörlin et al. 2022). Reterritorializing this Arctic region as part of the Swedish metal extraction industry connected Kengis and the surrounding areas, including the Tornio River and Lappea village, an emerging rhizome of the colonial global economy and practices (Nordin and Ojala 2017; Evans and Rydén 2007).

The expansion of the Swedish iron industry was intertwined with the economic imperatives and arrangements of the time. The Thirty Years' War (1618–1648) against the Holy Roman Empire taxed the Swedish economy heavily, and King Gustav II Adolf (1611–1632), needing to develop and mobilize new economic and financial resources, turned to the "great capitalists" of the time: the Walloon Calvinist investors and industrialists. He hired them to develop the Swedish metal industry and trade (Trevor-Roper 1980: 6–9). They heralded the rise of the capitalist system in Sweden, including Kengis. Furthermore, at the time of growing European colonialism and global trade, this arrangement connected Arctic metal production at Kengis with Atlantic colonial trade, including the slave economy (see Müller 2005).

Two Walloon industrialists, the Momma Brothers, were key players in connecting Kengis ironworks to this colonial rhizome of the global metal trade. Famous entrepreneurs of their time, the brothers were born in Aachen, in present-day Germany, into a family of copper producers with extensive trade connections to England, the Netherlands, the Americas, as well as networks in the Baltics, West Africa, and East Asia (Nordin and Ojala 2017:108–109). The brothers made their fortunes producing copper, iron, and brass, and while they invested in several iron mines and iron ore refineries in Sweden, the Kengis ironworks in the Arctic played a special role in their business and political aspirations.

Using their capital and connections, the brothers joined Sweden's political and economic elite and became the personal financiers of Queen Christina (1632–1654). The brothers were ennobled in 1669 for services to the crown. As part of their transformation from Dutch entrepreneurs to Swedish nobles, the Momma brothers assembled Arctic land, an Arctic manor, and Arctic codes to create a perfect milieu for their Swedish noble life. They reterritorialized themselves to the Swedish Arctic and constructed it as the territory for their new noble life. The Arctic region was available for this kind of reterritorialization, given its image among the elite as an uncharted frontier and thus not claimed yet by any of Sweden's noble houses.

The Momma brothers' reterritorialization project of becoming a noble Swedish house involved several strategic discursive recodings of their connections to the Arctic. They used reindeer, a half-tame species of deer that lives in the Arctic region, as an Arctic anchor. For their new noble family name, they invented the Swedish name Reenstierna, which means a reindeer star, also an invented celestial object. The brothers circulated an image of a reindeer as their noble house index, depicting reindeer antlers in their arms and a running reindeer on the coins they were given the privilege to manufacture at the Kengis ironworks. Their most impressive Arctic staging was the Kengis manor, the northernmost, grandest Arctic manor in the region, which they built as part of the Kengis ironworks. Similar to other Dutch entrepreneurs, the Momma brothers bought great estates in Sweden, a material index of their social status. The brothers owned several properties across Sweden but used the Kengis manor to reterritorialize themselves on Swedish Arctic territory in the process of becoming a Swedish noble family (Naum 2018; Nordin and Ojala 2017).

The Kengis manor soon became regarded as an outpost of modernity in the colonial landscape of the time. The manor, something not seen before or since in the region, was three stories tall, with twenty fireplaces, stained-glass windows, a Baroque garden, and an entrance avenue (Norden 2020). All these were material codes for progress, power, and prosperity. The manor was designed more as a political and economic formation than a family house. Still today, the Kengis manor is the northernmost manor in the region.

The Momma brothers never lived there for long but instead used the manor for business. The manor became world renowned through the works of famous Arctic explorers, who visited the manor en route to the north. In his book for 1877, an American author, Bayard Taylor, described Kengis Manor as "an oasis in the Arctic desolation" where one can get anything, including a decent bed, a proper dinner with wine, and civilized conversation. Kengis became a welcome stop for several explorers heading north, including French explorers and authors Regnard (1681) and Mermier (1842), Swedish scientist Carl Linnaeus (1732), and the first Finnish ethnographer Paulaharju (1923–1924). The accounts of these visitors to the Arctic and its people impacted for centuries on how the region and locals were perceived by outsiders (Isaksson 2001).

The *Momma* brothers created a new economic assemblage to transform Kengis into a profitable iron ore production site under emerging capitalist conditions. This meant assembling such material and discursive elements that together would produce the desired outcome: an economically successful ironworks. While Kengis already had the natural resources needed for iron production, it lacked capital, skilled labour, and governance fit for economic production. These were imported by the Momma brothers.

The core activity of Kengis ironworks was to process iron and copper, brought in from several nearby mines. The geographical location was ideal for the ironworks. The location by the powerful Tornio River and Lappea rapids provided both energy and a water highway for goods and people, connecting the ironworks to the Gulf of Finland and, ultimately, the Atlantic (Nordin and Ojala 2017).

In addition, the vast surrounding forests guaranteed a constant supply of the timber needed to produce coal for the metal production process. Finally, from the capitalist perspective, the locals provided a form of surplus labour, trainable for the mining work. To add value to the raw materials, the Momma brothers imported new technology, skilled French-speaking iron workers from Walloon (current day Belgium), and the capital and networks to participate in global trade (Müller 2005; Rydén 2002; Vilkuna 1996). The reterritorialization of new bodies with new languages, skills, and practices, together with capitalist and colonial regimes practiced by the Momma brothers, reconfigured the relations between people, resources, and the place. It turned Kengis and the surrounding villages into economic space with a new governance practice. The capitalist assemblage was ready to function.

Kengis Ironworks became a busy town in its own right. As a favour to the Momma brothers, the crown granted several exceptions to Kengis ironworks, each contributing to its economic success. The ironworks was granted town-like rights for trade and the extremely rare right to produce its own currency, which was needed for trade in a region that lacked cash. The crown also exempted the Kengis Ironworks from paying a toll to the capital Stockholm. Consequently, the ships carrying iron from Kengis could directly enter the global market via Amsterdam, where another Momma brother worked (Nordin and Ojala 2017: 113–114). Under these conditions, Kengis prospered. With town-like privileges, Kengis became a centre for trade between Finnish, Sámi, and Swedish locals. The famous annual Kengis trade market attracted tradesmen from far, even from Karelia in Russia (Paulaharju 1923).

The economic impact of the ironworks travelled beyond *Kengis* village. The surrounding villages, like Lappea, where Pentti and I live as part-time neighbours, provided not just the workforce but also food, timber, and coal for the ironworks. The main product was coal, which villagers produced by burning timber in special coal pits. Today, the remains of the coal pits can be found alongside the village's east side of the Tornio riverbank. One crispy autumn morning, Pentti took me to see the pits opposite my husband's grandparents' house. I was only able to see a very slight hill covered by blueberries and surrounded by tall pine trees. As a fieldworker for a museum association, Pentti had mapped the pit remains in the municipality, and they are now listed as protected relics of the region's early industrialization. There are also other traces of the economic impact of the ironworks. For example, some of the best fields and meadows, reserved for growing hay for cows that once produced milk for the ironworks, go by names like *papinniitty* (the meadow of the vicar) or *piimäkaltio* (buttermilk spring), indexing the historical function of these fields. Over time, the mobility of people and products between Kengis and Lappea and other Arctic villages became so frequent that a rhizome of paths emerged, connecting Kengis and the villages to Lapland's large inland towns and all the way to the Ice Sea. As Pentti told me several times when we were fishing from his river boat at Tornio River, what is now a peripheral village was once at the centre of early modern capitalist development in the Arctic.

3.3 In the Name of Progress: Creation of Mining Governmentality

Fig. 3.3 The water power for the ironworks

Despite all the investments and privileges, Kengis ironworks was never economically profitable. The capitalist mining assemblage did not function properly under the Arctic conditions. The harsh Arctic environment meant that the ironworks could operate only a few months of the year, and the transporting of products proved to be both costly and precarious (Komu 2020: 53). Instead, Kengis became a laboratory of early modernity in the Arctic through connections with settler colonialism and the emerging capitalism system (Naum 2018; Nordin and Ojala 2017; Nordin 2020). It transformed the Kengis ironworks into a colonial, economic space that relied on the extraction of local resources for global markets elsewhere while erasing the governance and presence of indigenous Sámi people and facilitating the control of the crown/state. The alliance between the capitalist mining assemblage and the political state assemblage strengthened (Fig. 3.3).

Kengis ironworks exhibited the hallmarks of modernity. It contributed to the transformation of the commons into commodities through investments, technology, and labour mobility beyond the region. In the progress of the deterritorializing of the region as indigenous land and reterritorilizing it into an Arctic extraction frontier, the ironworks drew on structures and practices of settler colonialism and consolidated enduring structures of inequality (Junka-Aikio 2023; Kuokkanen 2022; Spangen et al. 2015). It imposed new spatial, temporal, and moral orders, which were embodied as relations of subordination and strengthened by the growing alliance between capitalism, the state, and the church. In Foucauldian terminology (Foucault 2007), a new kind of governmentality was created, one that was not only concerned with maintaining order and enforcing laws but also creating mentalities, rationalities, and techniques to manage and control the people living and working for the ironworks (Bell 2023; Scott 2005; Schoderer 2023). The architecture for colonial and capital extraction was put in place.

Complex reterritorialization processes of the capitalist and colonial structures and practices reconfigurated and recoded the relationships between people and land (Deleuze and Guattari 1972/1977). The life and social organization of the indigenous Sámi people became deeply disrupted and dislocated. Their rights to the land were erasured, and the indigenous traditional land became dispossessed through a gradual process of recoding these lands as "unoccupied" frontiers, free to be given to the settlers by the state or to be claimed by the state (Spangen et al. 2015; Tuori 2015). As a discursive strategy, the Sámi people were recoded with a colonial category of nomads without rights or ownership of the land. This coding had an extensive impact on Sámi political, cultural, and economic position and rights. Still today, the Sámi people struggle to regain access to the land and natural resources. (Junka-Aikio 2023; Kuokkanen 2019; Valkonen et al. 2022).

Kengis ironworks became a capitalist assemblage supported by state politics. The Swedish name, *Kengis bruk (pruukki in local dialect/ Meänkieli),* already captures the core idea of capitalist production mode in the ironworks. Nordin and Ojala (2017) explain how the Swedish verb *bruka* literally means both to "to cultivate" and "to use". It is related to the words "to plant" and "a plantation". The two forms of production are related. A *bruk* is a single-purpose, concentrated production unit, just like a plantation, intending to extract and refine natural resources, usually metals (Tsing 2016). The *Bruk* is run by an intense labour concentration, living on the production site under strict rules. In the seventeenth- and eighteenth-centuries Swedish Arctic, the *bruk* (in the following called "works") signified a particular type of organization of resources, work, labour, time, and space, which was regulated in alignment with reformed Christian ethics, Dutch aesthetics, and early capitalist modes of production (Nordin and Ojala 2017).

This mode of production required new kinds of governance techniques. As Foucault (1978/1991) powerfully argues, modernity relies on governance that is designed to shape, regulate, and manage the conduct of individuals toward specific ends. To refer to this *conduct of conduct*, Foucault (2007) coined the term governmentality to underline that this new *art of governance* operates through producing governable subjects whose behaviour can be directed in a certain way. Governmentality creates rationales, practices, and norms that appear as the natural order of things, masking their ideological investments (Scott 2005: 201). It produces embodied relations of hierarchy and subordination, aligning with the capitalist production mode. The new governmentality was gradually distilled through discourses and practices familiar from extraction work in plantations and early industrial sites in other colonial contexts

(Naum 2018; Nordin 2015). It turned Kengis ironworks into a capitalist mining assemblage, operating within the lines of colonial logic and producing an emerging class system connected to the redistribution of resources and profits.

At Kengis Ironworks, this emerging governmentality was coded as "progress". It relied on three main strategies: labour importation; establishing an industrial spatial and temporal order; and implementing moral order through a strong alliance with the church.

The labour importation made Kengis ironworks the region's northernmost hub for labour mobility. The imported workers were the Walloon ironsmiths, who were considered the best in the world at the time. They were essential for the new technology, known as the Walloon-forge, which improved the quality of the process and products and was one of the Momma brothers' major investments at Kengis. Similar to other mining investors at the time, the Momma brothers also contracted Walloon ironworkers from their native Walloon regions for their mines in Sweden. An estimated 1000–1200 Walloons arrived in Sweden in the seventeenth century, which meant that one-third of the mine workforce was German or Walloon (Vilkuna 1996).

Several Walloon families arrived also in the Kengis area. Labour contracts were long, and French-speaking Walloons came with their families, who developed ties in the area and, over time, married locals. Scarse historical writings available point to a gradual language shift that took place over two generations. A French explorer and writer, Jean Francois Regnard, visited Kengis in 1681 and described his encounter with a Walloon ironworker, Pierre Servais, who had left Namur, the capital city of the Wallonia region of what is Belgium today, as a boy with his parents, Jean and Sylvianne Servais. Pierre Servais started working at *Kengis* Ironworks in 1654 as an ironsmith. Regnard describes the encounter in his travel book:

> We were delighted when we found out that there was a Frenchman in the area. You see, gentlemen, there is not a single place where the French have not found their way. Then he worked in the mine for 30 years, and sure enough, he looked more like a wild man than an ordinary man. He helped us with many things, even though he had almost forgotten

> his language. He assured us that in the time he had been in that region, he had never seen a Frenchman or met a closer neighbor than the Italian who passed by some 14 years ago and was never heard of again. (Probably Francesco Negri, who visited Svappavaara in 1663). We tried to find out about the man in his mother tongue, at least a little and learned many things from him that we would have found difficult to find out from someone other than a Frenchman (Regnard 1968/1731).

Locally, Pierre Servais became known as "the Frenchman", presumably referring to his French-speaking family. The Finnish version of his name was *Petter Servio* and later *Sirviö,* and he is considered the ancestor of the large family of Sirviö that has thousands of descendants in the Tornio Valley and beyond. Given the small population that lived in the area, most local families, such as my husband's, claim that they have "French heritage". When the Finnish ethnographer Samuli Paulaharju came to Kengis and Lappea villages three hundred years later (1920–1923), the stories about French-speaking, dark-haired Walloon workers still circulated (Paulaharju 1923: 72).

Labour mobility turned Kengis into a multilingual community where Finnish and Sámi speaking locals encountered Swedish, Dutch, French, and German-speaking explorers, industrialists, ironworkers, and traders (Nurmi 2019: 100; Nordin and Ojala 2017: 109). In the early mining economy context, language indexed not only ethnicity but also emerging class divisions and social hierarchies. The first French-speaking Walloons worked mainly as highly valued smiths, experts in a craft that was handed down from father to son. Swedish speakers at Kengis were mainly people from southern Sweden who were working in administrative positions. For example, the Momma brothers contracted a priest, a schoolmaster, a doctor, and a barber to work at Kengis (Fagervall et al. 2006: 20). Finnish-speaking locals worked mainly in the mines, but also in transportation and services needed for the mine operations, including timber and food production (Komu 2020: 53). The Sámi were essential for the transportation of ore in the early mining projects since hundreds of reindeers were used for transportation, and only the Sámi had reindeer sleds and knowhow in traveling in the Arctic environment (Komu 2020: 53; Ojala and Nordin 2015: 12) On the Kengis payroll, Sámi, Finnish,

Swedish, and French were categorized in separate groups (Kuoksu 2003; Ojala and Nordin 2015:120), but we can assume that many local workers were multilingual, at least between different local languages. Historically, flexible multilingualism was more of a norm than an exception in the region (Pietikänen et al., 2010).

The mobile, multilingual workforce needed to be shaped and conformed into docile, productive mining workers to make the mining governmentality work. *Kengis* became a site for shaping locals into docile workers and decent citizens. To accomplish this, the *Momma* brothers reassembled their mining assemblage to include elements of state assemblage politics and implemented three powerful governance strategies: spatial governance, the installation of industrial time, and the implementation of a religious moral order.

The spatial governance of the mine resembles a landscape of profit and power. Similar to other economic sites, such as plantations (Singleton 2001), or factory farms (Tsing 2016), in Kengis ironworks, buildings, roads, housing, iron production facilities, and sanitary conditions were designed to manage, control, and surveil the production to maximize profit. This spatial governance logic also reinforced hierarchical power relations on site. Writing about governance strategies, Foucault (1975/1977) argues that discipline "works best in an empty, artificial space that has to be completely constructed". Kengis Ironworks can be seen as just such an artificial space, a reterritorialized construction of modernity operating through particular rules and logics. Spatially, the Kengis ironworks was well-defined and territorially bounded, thus controllable with relative ease (Naum 2018: 13).

The spatial governance of Kengis ironworks was visualized in 1660 in a drawing below made by a Walloon worker Denis Joris (Fig. 3.4). It was most likely commissioned by the *Momma* Brothers as, at the time, paintings of factories and depictions of economic landscapes were often commissioned by investors hoping to brand themselves and impress their international trading networks (Nordin and Ojala 2017: 118). Although detailed archaeological studies have not as yet focused on Kengis ironworks, based on studies of nearby ironworks and the site's remaining buildings, the drawing accurately illustrates the ironworks in the Momma brothers' day (see Naum 2018; Nurmi 2019).

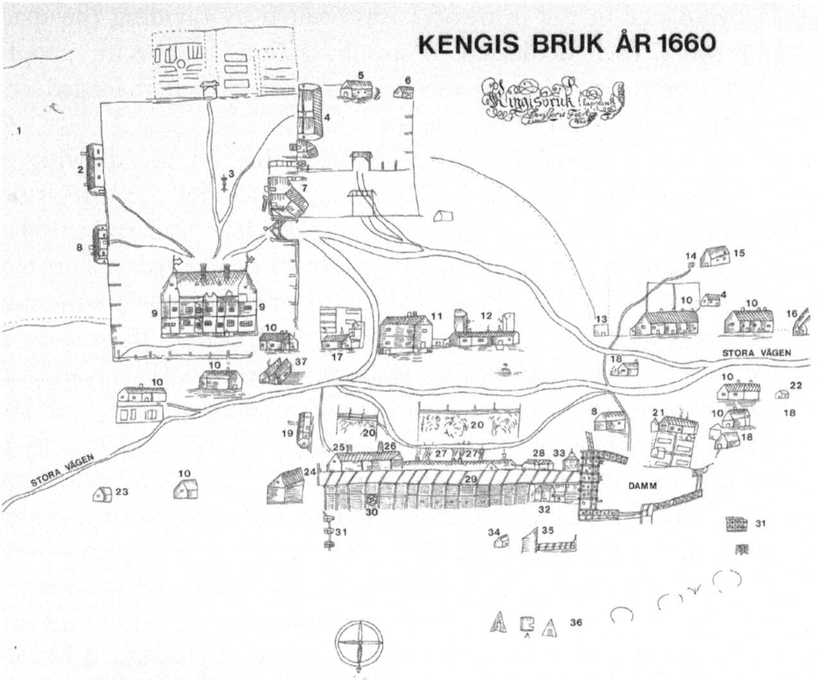

Fig. 3.4 Spatial governance of Kengis Ironworks 1660 (*Source* Jernkontoret)

The drawing depicts an early capitalist and colonial landscape: reorganization and restructuring of the mining site for economic purposes with emerging class and ethnic relations. The material reterritorialization of power relations and hierarchy is visible in the drawing, especially regarding the housing arrangements. Investors and skilled workers, i.e., the Walloons and the Swedes, lived in houses with chimneys, while Finnish workers resided in smaller houses with small windows and no chimneys. The Sámi inhabited their traditional *goahti* (a hut) dwellings on a small island separate from the main *Kengis* premises.

At the time of the drawing, Kengis was a community of 240 people (Fagervall et al. 2006). The site was divided into a hierarchical production space, resembling Foucault's "strict spatial partitioning" (1975/1995: 195) as an early form of panopticism (Sioh 2010: 470), where everything and everyone has a designated place and could be observed.

Spatial governance in the ironworks was created by dividing the space into four zones, each dedicated to specific actions by specific people. These zones of action were connected by roads, which managed and controlled the movement of the bodies.

The first economic zone, on the left-hand side of the drawing, is a gated space of power and administration. It includes the three-story renaissance manor with two chimneys, two porches, lead stained-glass windows with shutters, and a brick roof (Nordin and Ojala 2017: 69). The manor also included a church hall. Next to the manor are the residences of the mine manager and the priest, and nearby, the houses of the most valued workers, including Walloon ironsmiths (Fagervall et al. 2006). The space is divided into geometrical parterres with a fenced and gated garden. A sundial in the middle of the courtyard was a novelty in the region (Nurmi 2019). The second zone on the right-hand side of the drawing, by the main road, depicts the lodgings of the Finnish workers. The houses are smaller than those in the first zone and do not have chimneys. Workers had access to a small slot of land for their self-subsistence, and it was usually the woman who tended to small-scale agriculture and animals (Fagervall et al. 2006). The third space by the Tornio River was dedicated to ironwork. Industrial production was located along the river and the stream as the main energy source. It was also at a distance from the settlements, given all the noise and smoke of the works. In the fourth zone, at the bottom of the drawing, outside the grid and depicted on the islet, lie two Sámi *goahti* (traditional huts), distinctive for their traditional triangle shape. There are two explanations for the isolation of Sámi locals. One is that the Sámi were not allowed to live among other workers but were expelled to an island to be monitored and their movements limited. Another explanation from some archaeological research suggests that the Sámi had been living on the islet for centuries before the iron works were established, and the Sámi were still remaining there at the time the drawing was made (Nordin and Ojala 2017: 69). The two explanations are not mutually exclusive, and perhaps both are valid explanations. The island was a historical settlement of the Sámi while new governance practices emerged, facilitating their surveillance. Under the new circumstances, the spatial location of Sámi set them apart from the rest of the community and placed them in the margins of power relations.

Another important resource for mining governance was the introduction of industrial time (Naum 2018; Nurmi 2019). The idea of controlled time and devices for it was imported from the southern mines. Industrial time organized the activities at the mine into work, breaks, and free time. Kengis Ironworks adopted clocks in the seventeenth century (Nurmi 2019: 112). The drawing above depicts a large sundial in the garden, and Kengis also had a sundial placed in the front garden for servants and inhabitants of the manor. An hourglass and a large clock were also listed in the inventories of Kengis ironworks. These tools for time measurement helped establish rhythms, maintain schedules, and regulate the cycles required for productivity at the mines (Naum 2018: 13). Time management was needed to function efficiently since a complex system of different mining processes in different locations involved a considerable amount of people working to synchronize their actions. The problem of synchronizing the activities in the north was a challenge for capitalist production as traditionally, locals functioned in terms of a nature-based rhythm related to the amount of sunlight, snow, etc. (Nurmi 2019: 110).

Kengis ironworks was an Arctic version of a disciplined space in the Foucauldian sense. The drawing above shows how the ironworks were imagined and managed as orderly and ordering settings with symmetries and hierarchies. It conveys a vision of productivity at the ironworks. Spatially enhanced hierarchy, order, and discipline were built into the mining landscape (Naum 2018: 16). The ethnic and linguistic division of space was a recurrent pattern in the eighteenth-century Arctic, visible also in churches and marketplaces (Nordin 2020; Ojala and Nordin 2019). Furthermore, the creation of this kind of economic, disciplined landscape was a form of an early panopticon and, as such, a form of reterritorialization of colonial and capital logic materialized in plantations.

The remaining major challenge for capitalist mining governance was to transform locals into docile and productive industrial workers. Among locals, mining work had a bad reputation: the work was difficult, even life-threatening, and the pay was paltry. Poor living conditions caused epidemics, and the foreman had extensive permissions to punish the workers for lateness, negligence, or drunkenness. There was very little free time, except Sundays (Fagervall et al. 2006: 8–29).

In May 1654, Abraham Momma wrote to the High Chancellor of Sweden, Axel Oxenstierna, that the greatest obstacle to developing the ironworks was the negative attitude of the local population, which manifested in an unwillingness to work or learn the ironsmith profession. Writing from his colonial perspective, Momma complained in his letter about the laziness and irrational lack of interest in earning an income among the local (Finnish, Sámi) peasants "who spend the entire winter sleeping in their houses and would rather hinder than support the works" (see Naum 2018). Momma's complaint offers an Arctic example of a widely documented racialized and colonial construction of "lazy natives," which was used to justify capitalist extraction and unequal power relations (Borah 2021). As elsewhere, in Kengis the answer was that the locals needed to be transformed—improved and civilized—and discursively, this was done with help of discourse of *progress*. The Kengis mining assemblage now had a new element.

The Arctic discourse of progress enacts a still-circulating colonial code about the blessings that Western civilization is supposed to bring to the land and the people living in it. The "gifts" of Western civilization and Christianity were used to legitimate extraction and exploitations, typically by creating racialized categories and hierarchies of governance (Naum 2018: 20). At Kengis Ironworks, the discourse of progress combined several political, economic, and religious interests and ideas that all aimed to produce a docile workforce for the mining industry. As at other colonial extraction sites, in Kengis, progress became intertwined with three powers: capital, the state, and the church. Investors wanted to reap profits, the state/crown sought to strengthen its hold on Arctic land and people, and the church aspired to expand its influence in the Arctic. The alliance between economic, political, and religious interests made Kengis a commercial and religious site too: Kengis became a town with a church, shops, a jail, and its own police force.

The Arctic discourse of progress, rooted in Enlightenment ideas of social and individual improvement, aligned with the colonial and capitalist ideologies in shaping locals as pious congregants, productive miners, and compliant citizens. In practice, this entailed not only training the locals in new skills and new ways of working but also shaping them into governable, docile subjects for an emerging capitalist system (Naum 2018: 7). This discourse reterritorialized religious, colonial, and capitalist values and practices at Kengis ironworks and recoded relations between resources, people, and power in terms of hierarchy drawing on an emerging class system. It intertwined piety and productivity to create an economic-religious code of conduct and sense of morality embedded in the new capitalist, class-based resource redistribution. A key discursive strategy was to create strong connections between prosperity, usefulness, and virtue, connecting godliness with diligence and pious and honourable behaviour according to the divine order and natural laws (Naum 2018). Becoming a good worker and a pliable citizen meant adopting improved attitudes towards industrial mining work and its governing techniques.

The discourse of progress regimented locals into pious mining bodies at a moment when the smooth Arctic space changed into a stratified space of capitalism and colonialism. It provided rationality and legitimacy for a new Christian moral governance practiced in Kengis. By the Queen's order, the Momma brothers built a church right next to the historical marketplace and hired a priest (Paulaharju 1923). The creation of moral order at the mines was reinforced by a special legal code issued by the Swedish Board of Mines in 1640 (Naum 2018: 13), which defined what was obligatory and what was forbidden at mining sites. Sunday church services and morning and evening prayer became obligatory at the ironworks. Abusive drunkenness and the absence of religious practices were forbidden and punished (Naum 2018: 13). The church's harsh missionary role became evident concerning indigenous Sámi people. Naum describes how all Sámi who did not attend church services and lacked knowledge of the Christian creed were forced to work until they showed "serious improvement and willingness to follow God's teachings and withdraw from the darkness of ignorance" (Naum 2018: 11–12). The church recruited Sámi speaking priests and hired interpreters with the goal of converting Sámi people to Christianity. Another technique was the creation of a rhizome of dependency, privilege, and punishment. Some key goods, like flour and tobacco, were only allowed to be sold to the workers at the ironworks. Offenses and criminal records could be amended by working in the mines and locals could be excluded from military service by working at the mines (Fagervall et al. 2006).

The new mining governance at Kengis ironworks paved the way for an expansion of Kengis production in the eighteenth century, alongside the development of the northern Swedish Kingdom. The crown actively promoted new settlement, agriculture, new livelihoods, and new living standards for Lapland locals (Haikola and Anshelm 2020; Naum 2018). Mining continued to be an important source of wealth for the state and investors, and increasing the northern population was thought to also be beneficial for exploiting the northern metals and minerals. It is hard to believe now, but in its heyday, in the middle of the eighteenth century, about 120 people worked at the Kengis ironworks, and about 500 people lived in the community. The owner at the time, Abraham Steinholtz, who bought the business from the Momma brothers, invested in and developed the ironworks in many ways, including adopting new refining techniques. After Steinholtz, the Solberg family acquired the ironworks, but industrial production was discontinued after the ironworks burned down in 1879. Today, the manor house continues to be the private home of the Solberg family, but instead of mining, the old ironworks grounds are now a fishing tourism site, connected to growing tourism industry in the region.

3.4 Radical Re-coding of Religion and Communism in the Mining Frontiers: Leastadius and Lenin in Mines

Fig. 3.5 The Kolari Church with the red cross

The nineteenth and twentieth centuries changed the Kengis and Kolari regions into the terrain of class tensions between workers and capitalists. In assemblage terms, the politics of capitalist assemblage become stronger while the politics of state assemblage was contested and fought over (Fig. 3.5). The time was one of political and social turbulence: it included the Great Famine, the independence of Finland, two world wars and a Civil War, the industrialization of Finland, the establishment of communist parties in Finland, and unionization. In Lapland, all these changes became intertwined with mines and consequently, the mining assemblage was reconfigured constantly. These historical developments explain at least partly why mining still continues to be part of the current political debates.

In the nineteenth century, state mining governance in Kengis, created to consolidate capitalist resource redistribution, was challenged by the most powerful and radical religious and political movements in the history of the Finnish Arctic. First of these nomadic movements was a pietistic Lutheran revival movement called *Leastadianism*, named after its leader, the preacher and priest of Kengis Ironworks, *Lars Leevi Laestadius*. Second was a local version of the communist labour movement, called "*korpi-kommunismi*" (wilderness or backwoods communism), which referred to the movement's peripheral geographical position and its organization around the industrialization of forestry (Aatsinki 2009). Both movements reconfigured mining assemblage by introducing a new moral and political order related to work, distribution of resources, and ideas of "progress," still palpable in current discussions on the value of mining work (see Sect. 3.4). These movements were part of profound changes in the Arctic region at the time, brought forth by the inequalities related to the capitalist form of production outlined above. Both movements reflect efforts by poor and working class people to cope with rapid changes in their environment (Aatsinki 2009), through the processes of de/reterritorialization and re-fitting their labour and bodies to changing economic and political assemblages. In the local context around Kengis iron works at the time, these poor, working-class people were mainly Finnish speaking.

The founding figure of the radical religious movement was the priest of the Kengis church, Sámi-Swedish Lars Levi Laeastadius (1800–1861), a trained botanist and academic author who spoke and wrote in several languages, including Southern Sámi, Pite Sámi and Northern Sámi, Finnish, Swedish, German, and Latin (Ojala 2016). As a young scholar, he worked for a French Arctic expedition, sent by King Luis Philippe I in 1838–1839. After his "awakening" in 1844, Laestadius focused on his religious calling and became a charismatic leader and preacher of the revivalist movement, with sin and repentance as a core doctrine. Laestadius became known for his strong, vivid preaching, and ecstatic sermons delivered in Sámi, Finnish, and Swedish. While gaining an increasing number of followers, referred to as "true believers", especially among local Sámi and Finnish people, he was a controversial figure, opposed by many powerful people and institutions, including the patron of the Kengis ironworks at the time, Karl Solberg, and some priests in the area (Ojala 2016). Leastadius and his movement were considered radical at the time for disturbing the existing order and power relations, for example, by inviting laypersons to become preachers, incorporating Sámi mythology and religion into mainstream Lutheran doctrine, and criticizing the authorities. He appealed to the local workers by preaching the chance for hope and redemption for all people who followed the movement, not just elites. A better (after)life was promised for true believers who held pietist values central to their lives, including sobriety. Progress would come through redemption and temperance. Towards the end of the nineteenth century, the movement started to lose its radical status when it began supporting more conservative values and encouraged conforming to centralized power structures (Nykänen and Linjakumpu 2020: 25–27). Today, a Christian Pietist denomination bears the name Leastadius and has an estimated 150 000–200 000 members, mainly in Finland. It continues to be influential in the Arctic region, for example, in terms of promoting traditional values and the Protestant work ethic (Nykänen and Linjakumpu 2020).

The rise of a radical religious movement was followed by a radical communist labour movement that continues to be influential in the Finnish Arctic. The emergence of the communist labour movement in the Kengis area was part of a complex rhizome of overlapping social and political changes. As in much of Europe, the Arctic labour movement was rooted in the tensions between capital and labour, enhanced by intensifying industrialization. Unlike its southern Finnish counterparts, which developed at textile factories that provided year-round employment in urban settings, the radical Lapland labour movement was spurred by inequalities experienced in the mines. At the turn of the twentieth century, these inequalities were further deepened by the rapidly intensifying industrialization and growing forestry industry that offered seasonal work only and with poor conditions (Aatsinki 2009). Lapland's great logging era extended from the 1870s to the machine-operated forestry of the 1960s and 1970s. Vast forest areas, Lapland rivers, and surplus Arctic labour bound the region to the industrializing capitalist economy, global trade, and an emerging division of labour and territory: the role of the north was to provide raw materials for industry and the European markets (Aatsinki 2017). This role fixed the region as a periphery of the developing nation-State of Finland and a resource reservoir for its developing economy.

The massive expansion of the forestry industry reinforced economic and political tensions between social classes and created a new kind of working class in the Arctic. Landless people who previously made their livelihoods within the networks of village and family ties became surplus bodies seeking wage-based work. The great Finnish famine of the 1860s paved the way for the radical communist movement. Nearly ten percent of the Finnish population died of hunger, and six percent left their homes to look for food (Nevala 2015). In Lapland, people moved not only to the south but also to North America and the shores of the Ice Sea in Norway, where they became part of what is now referred to as the Kven minority (Lane 2016). Those who remained suffered three years of hunger. Meagre support, such as flour and grains sent by the state, was distributed by the bourgeoisie, but there were claims that this support was not evenly distributed (Nevala 2015).

When the big Lapland logging, the start of the industrialization of forestry, began a decade later, it seemed like a lifeline for many locals. For the next hundred years, thousands of Laplanders got a significant portion of their daily bread from the region's logging industry. In addition, new labour began to flow into Lapland as the forest industry expanded its operations from the 1890s onwards. Lapland's population doubled in just three decades (Aatsinki 2019). One of the largest forest companies operating in Finnish Lapland was *Kemi OY*, which in 1910 employed an estimated 10,000 workers from the Tornio River and Kemi River valleys alone (Snellman 2005). Men from the Kolari area and Lappea village also worked for the Kemi company, especially drifting timber along the Tornio River, as in the early twentieth century, rivers were still the main highways for transporting goods (Aatsinki 2019).

While the forestry industry provided possibilities for locals to earn income, it also created an Arctic precariat. Forestry work was hard labour, and much of it, such as cutting and timber drafting, was seasonal, usually lasting only three to four months. The work was not just physically demanding but also dangerous. Landless men mainly did timber rafting since the work did not require expensive equipment. At the same time, the timber drifting and especially the explosion of the riverbed to better facilitate rafting endangered the traditional livelihood of the Tornio River, namely salmon fishing. Pentti explains that several traditional salmon fishing practices, such as fish-trapping, were no longer possible along the exploded riverbanks, and several, so far, well-off families in Lappea have become poorer. The company also "blew up the river" to clear out the big stones and deepen it to better accommodate timber drifting. This had a devastating impact on the ecology of the river and its fish in particular. Several owners of the big Lappea houses had to leave for forestry work. Pentti's father and uncle worked in timber drifting for several seasons along the Tornio River. One summer, his uncle drowned during the timber rafting. Even today, when Pentti remembers his childhood and the economic changes the forestry industry brought, he becomes bitter. He describes the actions of the big timber company in the Tornio River valley as "a scam" and "a robbery", referring to the company's practice of setting up a local office and persuading local people to take larger loans than they could handle. Local people's rights to the river and forest land ended up in the hands of the company. "They robbed the whole Tornio river valley", Pentti summarizes.

The forest companies also earned a bad reputation in other ways. They opened shops at their worksites where the workers could buy food, but because of the monopoly, prices were high, and the food quality was often poor. There were no medical care and no restrictions on working hours. The companies did not support worker unionization and stopped strikes with violence. In addition, the foremen of the forest companies reportedly ruled sites with an iron fist, an approach that was amplified by class differences. These translated to differences in political alliances: managers supported right-wing politics, while workers supported the communist parties (Aatsinki 2019). A growing number of working-class people in Lapland started to identify with the goals of the labour movement, which advocated, among other things, universal and equal suffrage, freedom of religion, an eight-hour working day, and free education (Aatsinki 2017: 144).

At this moment of industrialization with growing class differences and identification, the local politics of territorial, state, capitalist, and nomadic assemblages were reconfigurated. The state assemblage increased its hold on Arctic land and forest by reterritorializing commons as state land and state forest. The state politics aligned with political right and capitalist interest in strengthening industrialization, while the workers, communist, and labour movements were seen as radical and nomadic, unsettling and challenging the state order. The tensions between the industrial mode of production and the aspirations of the workers for their rights continued to be particularly heated in Kolari, the neighbouring municipality of Kengis mine, located just on the other side of the Tornio River. Kolari was said to be "living between bread and meat", which referred to its marginal position concerning livelihoods: it was too far north for profitable crop growing and too far south for profitable reindeer herding (Niittyranta 2004). Thus, no bread, no meat. As Kengis mining work dried up in the late twentieth century, the forest industry seemed to provide opportunities for the Kolari people to earn a living.

Kolari became a nomadic hotspot of the radical communist labour movement in Lapland, where it was regarded as a way to impact the ongoing social change and win rights for workers. Throughout the early twentieth century, there had been active communication between Lapland and Russian communists (Tasala 2000). During the First World War, many prominent communist leaders of the time, including Vladimir Lenin, travelled through Lapland since northern train connections to Europe were considered safer than the southern routes. On his route back from Switzerland to Russia in April 1917, Lenin gave a speech to locals at Tornio railway station, encouraging them to revolution. Eight months later, in 6th of December 1917, Finland declared its independence from Russia.

Less than two months later, on the 27th of January 1918, tensions between the communist working class and the conservative bourgeoisie class escalated in the Finnish Civil War. This was waged between the forces of the Social Democrats, led by the People's Deputation of Finland, commonly called the "Reds" (*punaiset*), and the forces of the non-socialist, conservative-led Senate, commonly called the "Whites" (*valkoiset*). The Reds were supported by Bolshevist Russia, while the Whites received military assistance from the German Empire (Mähkä 2018). The short (only three and a half months) but brutal Civil War claimed the lives of 40,000 people out of a population of three million, and it left the country divided into two opposing camps.

During the Civil War, Lapland was a stronghold of the Red Forces, but the White Forces quickly and brutally took it. Aatsinki (2020: 54) argues that the White forces cleansed Lapland of potential revolutionaries: Lapland was the first place in Finland where the Whites exercised their execution and field court practice: 82 people were executed, and more than 300 were sentenced to prison camps. In Kolari, as elsewhere in Finland, the Civil War divided the community between "Reds" and "Whites". The following decades were characterized by political unrest and violence.

Aatsinki (2020) argues that White Finland moved from the control of space to the control of action in Lapland, and as a result, the Communist Party went underground temporarily. Nevertheless, the Communist Party and labour organizations continued to have strong support throughout Lapland. Before World War II, the communist parties received nearly 70% support in all elections in Lapland. Kolari became known as the "reddest" county in the country, and the left-wing parties were successful in Kolari until the 2021 elections.

Occasionally, political tensions in Lapland between the "Reds" and the "Whites" broke out into violent conflicts (Tasala 2000: 22–23). One of the most famous in Kolari took place in 1933 when a well-known local member of the White Guards shot dead a well-known local communist in front of his home. The local community knew both men, and both had family members and friends in both camps. Three generations later, the memory of this event continues to live on: I was told the story many times, offered as an explanation for the continuing popularity of the Communist Party in the region and the tensions between certain families and villages in the region.

World War II brought total devastation to Lapland, turning it into Finland's poorest area. Finland fought against the Soviet Union (Winter War 1939–1940 and Continuation War 1941–1944) and became an ally of Nazi German from 1941–1944. The devastation for Kolari was brought by two turning points in the war. The first was related to the violence between local communists and the ruling Whites in 1941/1942, as the former refused to take up arms in the continuation of the war against the Soviet Union. The second was related to the so-called Lapland War (1944–1945) between the former allies, German troops, and the Finnish army. Both developments scarred Kolari for decades to come.

For many communists in Lapland, participating in the Continuation War (1941–1944), which was seen as a war of aggression, went against their beliefs, and they refused to take up arms. The State of Finland interpreted this as treason with capital punishment. In the Kolari area, around forty to fifty men went into hiding in the vast forests of Kolari and Kengis, earning the nickname of "cone guards or forest guards" (Tasala 2000: 13). The forest guard impacted the whole community given that their fate lay in the hands of the locals, either in terms of help (providing food, clothing, hiding places) or in terms of being caught. The hiding men were hunted by the military, police, and local Kolari members of the White Guards, and their family members were questioned harshly (Tasala 2000: 14). The majority of the men were arrested in the fall of 1941 and winter 1942 on the Swedish side of the Tornio River. The Kolari arrests were violent: a state policeman was injured, three communists were shot dead, and one died during police questioning. The four communists were buried in the graveyard of their home village, next to Lappea village, where their tombstone reads: "Here rest the fearless young men, tortured and murdered by fascist authorities, who, refusing to fight a criminal war, fought for peace and democracy". After the war, in 1946, a military court and then two years later, the High Court of Finland convicted the men who took part in the arrest for violence (Tasala 2000).

The second wave of devastation was caused by the Lapland War between the former allies, German troops, and the Finnish Army. Following the signing of the Moscow Armistice with the Allied Powers, which stipulated the expulsion of German forces from Finnish territory, Finnish groups fought against the German groups from the fall of 1944 to the spring of 1945. The Lapland War left the region destroyed. As Germany retreated west to Norway, its armies adopted a scorched-earth tactic, bombing and burning vast areas of Lapland. An estimated 70–90% of the buildings, roads, and bridges were destroyed (Tuominen 2003). On their way to Norway, German troops came through Kolari and burned it to the ground (Niittyranta 2004). Even as late as the 1990s, when my parents-in-law started to build a summer cottage in Lappea village, they found an unexploded bomb buried in the field next to the village road.

Over 75% (more than 100,000) of the inhabitants of Lapland were evacuated, mainly to Sweden (Tuominen 2003). My mother-in-law, at six years of age, fled, with her pregnant mother and two younger siblings, to Sweden in the fall of 1944, crossing the Tornio River to the grounds of Kengis manor. They spent the winter in Sweden and returned in the summer of 1945 to Lappea village, which had been completely destroyed: all houses had been burned down, and the roads and bridges had been exploded. They first lived in state-provided barracks with other villagers until they were able to gradually build new houses for their families. The father of the family had died in the war, leaving the mother to provide for the children. She did that by working as a sewer. This was only possible because of her close connections to the other side of the river: her own mother was from Sweden and was therefore able to provide sewing materials that were non-existent in the devastated Finland back then. The experience of my mother-in-law illustrates the enduring and vital connections and exchange across the Tornio river, that connects the people living on the both side of the river (Fig. 3.6).

For Lapland and all of Finland, the postwar decades brought massive reconstruction and intense industrialization. During the reconstruction years, the political and capitalist assemblages became tightly interwoven in Kolari: it became both the "reddest" and the most industrialized municipality in the country per capita (Niittyranta 2004). Even the cross

Fig. 3.6 Destroyed Kolari Town, April 24.10.1944 (*Source* SA-Arkisto)

of the new Kolari church, built in 1965, shines bright red over the town. The ecclesiastical explanation that the colour indexes the blood of Christ failed to convince locals who saw the community's political colour reflected in the cross.

3.5 Power of Nostalgia: Living Mining Dys/Utopias

Fig. 3.7 Ruins of a mining village

The twenty-first-century Kolari is still a divided community. Deep-running historical rifts and alliances have surfaced regarding what to do with Kolari's closed *Rautuvaara/Hannukainen* iron ore mine, which is currently the object of intensive re-opening and extension plans. The mine has experienced boom, bust, and buzz cycles. As early as the sixteenth century, when the Kengis ironworks was established, attempts to extract iron ore from the Rautuvaara deposit took place, but the distance between the mining site and the ironworks by the river proved too difficult. Over 400 years later, the Finnish state-owned mining company *Rautaruukki Ltd*. opened two mines in Kolari: the underground *Rautuvaara* mine (1962–1988) and the open-pit *Hannukainen* mine (1978–1990). This provided year-round industrial work for approximately 200–250 people (Komu 2019: 54) (Fig. 3.7).

The two mining assemblage politics, state and capitalist, come together under the conditions of state mining capitalism (Kröger 2016), which refers to state politics where the state heavily subsidizes often unprofitable industry for seeking to keep the peripheries populated by providing jobs. This applies to Kolari, too. With state support, Kolari became a booming mining town in the 1960s-1980s. In addition to the iron ore mines, a state-owned limestone mine also operated at the same time (1968–1989), providing jobs for some 140 people and ten subcontractors (Salokorpi 2007). Kolari was an example of the economic and social politics of the State of Finland at the time. Large timber and mining factories turned many Lapland towns almost entirely into company towns, heavily dependent not only on one sector but on one particular company.

Kolari's golden industrial era began. The strong mining industry in Kolari became coded as a political and economic opportunity for the working class, and at this moment, thus far nomadic and radical aspects of the communist movements becomes, at least partly, reterritorialized as part of capitalist and state mining assemblages. The mining industry provided locals with much-needed wage-based work and contributed to the development of the region's infrastructure.

A key figure behind industrialization and leftist politics was *Eemil Lakkala*, the long-standing president and member of the Kolari Municipality Council (1945–1972) and a Finnish communist party MP. As a young man, Lakkala worked for the forest industry and helped lead Kolari's Communist Party. He became famous for his speeches on better rights for the workers and for organizing the local labour movement. Due to his political activism, Lakkala went underground and became a member of the Kolari forest guard, but he was eventually arrested and sentenced to forced labour camps several times between 1930 and 1945 (Tasala 2000). After World War Two, Lakkala became Kolari's leading politician, earning significant political power through decades-long work for labour unions and labour rights. With the support of labour unions and left-wing parties, Lakkala helped revitalize the mining industry in Kolari and opened two local mines. Mining in Kolari came to represent both economic and political opportunities for the working class. The mining industry provided locals with much-needed wage-based work and contributed to the development of the town's infrastructure.

The mining boom accelerated the shift from small-scale agriculture and seasonal industrial work to a fully-fledged industrial and service-based economy in just a few decades. For Kolari, the impact of this change was profound: it reconfigured class-based relations and political alliances. On the one hand, this was the golden age for the labour unions and the Communist Party. On the other, the increased prosperity, together with modern, even urban lifestyles, created middle-class aspirations and practices in Kolari, reflected in growing support to right-wing parties.

The mining boom created new forms of infrastructure for mobility and consumption. Even today, Finland's northernmost railway ends at Kolari, which sits at the crossroads of several important routes between north and south, and Finland and Sweden. Infrastructure for services and consumption expanded. Several banks and grocery shops opened in the town, as well as several specialty shops, including clothing stores and beauty parlours. *Taimi,* now a 60-plus pensioner with a working history of almost 20 years in administration in a Kolari mine, reminisced about a prosperous boom town and the amenities. Taimi fondly recalled the new—and only—hotel in the town, with white tablecloths, where it was "so fancy to go there for dinner, dressed up and everything".

The urban lifestyle was not limited to new opportunities for consumption but also gave rise to new independent housing choices for young people whose only option up to then had been to live with their parents or in-laws. The Kolari mines built housing for workers, including blocks of flats, semi-detached houses, and single-family homes. The mine created a bustling village around it, including housing for the workers and designated areas for social events and leisure, such as a clubhouse, soccer field, ice hockey rink, and vegetable garden. Similar to Kengis Ironworks, this mining site was also a hierarchal, organized space designed to manage and govern both work and workers. The mine's CEO enjoyed a large house, separate from other housing, while the management lived in semi-detached houses "with a view of the river," as Pentti described it. Workers like Pentti lived in a block of flats only "five minutes" from work.

The moral order was observed, too. Taimi tells a story about how she and Simo, her boyfriend at the time, were looking for company housing, but the answer was that the housing would become available only once they got married. Taimi and Simo married within two months and settled for ten years in the company housing, raising a family with kids in the company village with both parents working at the mine.

The temporality of regulated mining work, organized in three shifts, created new freedoms and tensions. Pentti recalls that even if you were only two minutes late to punch your timecard when your shift started, it was deducted from your salary. Regulated industrial time created a clear division between work and leisure, a new concept for many locals at the time. After work, instead of farm chores, mine workers had free time, and soon, various social activities were organized. Taimi explains: "We had to organize something as this was a small community, there was nothing there, and we were young people. We had to organize some entertainment for ourselves. So, we organized sports. The company supported sports a lot. The company also supported the women's club. And there was a clubhouse, too, where we were allowed to organize various events. So, there we lived, and there were our hobbies, and then the company even made us a vegetable garden". While Taimi fondly remembers this closely-knit community, "We were really like one big family," Pentti describes it differently, as "the time of Patron, a hierarchical system that cared for and controlled its workers".

The industrial time became an object of management and political tensions between workers and managers (Pietikäinen and Allan 2020). This was also the case in the Kolari mines where regulated working time created political tension between workers and the management. Like in all of Lapland, workers at the Kolari mines were highly unionized. The political dimension of labour relations was present in everyday mine management as unionized workers and their rights were closely monitored and defended. Simo, who worked as a middle manager at the mine, explained that he needed to pay careful attention to issues that might be easily politized, as there was always the possibility of a strike. He tells a story about a timecard to illustrate the tensions between workers and their rights on the one hand and managers striving for maximum productivity on the other. He worked the morning shift at the mine, starting at

7.00 am. If a worker was late, even one minute late, the time was taken away from their salary. One manager insisted that workers need to clock their 12-minute coffee break. The workers considered this unreasonable, given that the cafeteria was some 50 meters from the production facilities, and simply walking back and forth would leave hardly any time for a break. After tense negotiations, with a strike card on the table, the manager backed down, and the workers were allowed to keep their break without clocking it.

The boom lasted almost three decades, but the bust took only two years. It came as a shock to locals, even if the Kolari mines followed the well-known cycle of mineral mines: A mine's lifecycle is typically calculated in decades, and the bust comes quickly due to volatile global market prices. Moreover, like Kengis, the Kolari mines were never economically profitable and were heavily subsidized by the state. The Kolari bust came in the late 1980s and was linked to wider political and economic changes that led Finland to join the EU in 1995. Many Lapland mines and timber factories closed in the 1980s and 1990s; some were bought by international investors, and a few later reopened.

In Kolari, the mining bust was devastating. When the mines closed in 1989, the vast majority of the workers were left unemployed, and the majority moved out of Kolari (Niittyranta 2004). The remains of the mines were left to ruin. The company village turned into a ghost village as the block of flats, and the production facilities were left empty. In the following years, many services were shuttered, and any remaining business moved to a nearby major highway, leaving the old Kolari town centre empty. Kolari town turned into the ghost town it is today. "Only the undertaker is busy", Taimi says.

The affective power of mines in Kolari is significant, and nostalgia for the mining boom is palpable. The glory days of mining remain a living memory, newly surfaced by the buzz surrounding potential plans to reopen the old Rautuvaara/Hannukainen iron ore mine (Suopajärvi 2015). After the mine was closed in 1989, there was one attempt to reopen. In 2005, a transnational company, *Northland Mines Oy*, prepared for mining operations pending the permit, but the company was declared bankrupt in 2014. The second attempt to reopen the mine is ongoing. In 2015, a local transportation and mining services company, *Tapojärvi Oy*, brought all the rights and exploration materials from Northland Mines Oy, and established a new mining company, *Hannukainen Mining Oy*. The licensing process is ongoing, but the mining company speculates on a 2026 opening, and that it will operate for around 20 years. At the same time, there is strong opposition against reopening the mine, mainly because of environmental reasons and risks that mining poses to the tourism industry and its hotspot, a nature and skiing resort, located only 8 kilometres from the planned mine (Hast 2021; Similä and Jokinen 2018).

Aspirations to reopen an old mine are an affective abyss. The reopening plans evoke strong opinions both for and against. The question divides families, villages, and the whole community.

As the decision about the future of the mine continues to be debated, Kolari lives in an intensive speculative mode. The speculations run high regarding the chances of the mining company being granted all necessary licenses. As in many traditional mining communities, in Kolari, the question is polarized around two fixed positions: pro-mine and anti-mine, which has also materialized in local politics. The municipal elections in the spring of 2021 were exceptional. For the first time in more than a century of elections, the winners were not the traditional political parties but two movements, pro-mine and anti-mine, with the pro-mining camp winning the majority by one seat. These opposing views occupy local politics at the municipal council, where many crucial decisions related to mining licensing and infrastructure investment are made.

Surveys conducted by different stakeholders, including the Kolari municipality, the *Hannukainen* mining company, and the *Finnish Nature Resource Institute*, indicate that the support for and opposition to the reopening of the mine breaks down along age, location, and previous/current occupation lines. Mining support is strongest in Kolari town and in those southern villages that were once the strongholds of the Communist Party and that have not benefitted from nature tourism developed in the region. In contrast, strong resistance to the mine comes from northern mountain villages that depend on nature tourism and locals engaging with reindeer herding and fishing. Another dividing line is between the older supporters of the mine and younger opponents. Seasonal locals, such as tourists and holiday homeowners, tend, unsurprisingly, to be against the mine (Jokinen 2019).

A local man interviewed on regional television news summarized the pro-mine viewpoint: "Our town is becoming deserted by businesses, and that's why I would like to see Kolari revitalised like it was in the 1980s when the mines were running. If we get a mine, we will get young people, jobs, and families with children to come here, the town of Kolari will be strengthened, the businesses will resume operations, and we will receive tax revenue for Kolari" (Yle Lappi 2021). The affective force of nostalgia erases the memories of the mining bust, establishing a cause-and-effect relationship between Kolari's wealth and an operational mine in town. The inevitable bust, the global vicissitudes of today's mining industry, environmental risks, and forty years of nature tourism development are disregarded. Mining is seen as the lifeline for the town's economy.

The affective forces behind the anti-mine view are concerns about the "pure" Arctic nature. The anti-mine movement in Kolari assemblages a wide range of people and interest groups, which are all dependent on nature and its purity in their lives and business: environmentalists, reindeer herders, nature tourism entrepreneurs, fishers, and second homeowners in the area (Jokinen 2019). This alliance is connected by the particular location of the Rautuvaara/Hannukainen mine. The area still belongs to the traditional reindeer herding area, even though it is not part of the legally recognized Sámi domicile area, which starts some 40 km to the north. The planned mine, with the largest open pit in Europe (10 km wide), is only 8 km from one of the biggest nature tourism and ski resorts in Lapland, which brings significant income to the municipality. One of Finland's biggest nature reservoir areas is only 10 kilometres away from the mining site. Furthermore, the mining area is next to the nature protection area and a small river connecting to the Muonio and Tornio Rivers. The mine is not located in the periphery but at the heart of several key economic and cultural practices, to which mining poses a risk. If the mines previously considered to bring progress, they now were seen causing irreversible harm.

The pro-mine vision is rooted in the past, fuelled by the nostalgia of the former mining boom, and it ignores the impact of the latest mining bust, which is still evident in the landscape. This vision is harboured especially by the more senior generation that lived and worked in the mine. As Taimi explains, "I cannot think of anybody who worked in the mines at the time opposing the reopening of the mine". The glory days of mining are a generational memory, shared by many of the region's seniors.

Mining futures in Kolari re-configures the historical discourse of progress into two updated versions. In the first version progress is constructed around the rationale of securing full-time, all-year-round mining work linked to the organized workforce and vitality of Kolari. Its logic is based on the argument that it will work again since it worked well in the past. Taimi captures this: "I do not understand how it could pollute now as it did not before". In the other version, the progress is discursively linked with question of sustainability, evoking accelerating climate change in the Arctic, nature values, and environmental risks. The anti-mine voices argue that the mine is a short-lived venture with a permanent impact on the environment, which is an important resource for the area's present and future economic development, and a value per se, aside altogether from global green arguments and imperatives. Opponents of the mine also say that it is inaccurate to talk about "re-reopening" as the new mine would be 10 times bigger than the earlier one and not owned by the state but by international conglomerates. The pro-mine position is rooted in nostalgia for mining work, as well as for imagining a future with "real work" and "full employment," while the anti-mine position focuses on the prospects of both mining (impact on the environment) as well as the future sustainable economic development of the area (berries, fish, tourism). The two temporalities and prospects do not interact easily.

3.6 Mining Harm and Hope

Fig. 3.8 A old open mine pit by a river

Current debates on the future of the mining industry in the Arctic can be seen as a struggle to reterritorialize both mining and nature and to reconfigure the relationship between them while imagining alternative futures for the Arctic (Tolvanen et al. 2019). The state and capitalist politics of mining assemblage organizes the relations between minerals, nature, and work with the logic of progress, which, in turn, is seen primarily as increased economic prosperity. In the case of Kolari, this assemblage is temporally located in the region's past mining boom, and the affective force of nostalgia is powerful. The assemblage attempts to keep the mining territorialized in the terrain of good work and economic prosperity. In contrast, a more critical, nomadic mining assemblage attempts to de-territorialize mining from the ideas of progress and prosperity and reterritorialize it in the terrain of ecological risks and sustainable development. The temporal location is the future, the future of Kolari, the Arctic, and the whole planet. New political projects require new visions of what is understood as progress (Fig. 3.8).

The critical assemblage approach helps understand what is at stake in the mining debates in Kolari. Kolari is a conjuncture of past, present, and future memories and speculation for economic development around mining. It continues to be the periphery of a nation-state with diminishing services, fewer inhabitants, jobs, working-aged people, and children. Some infrastructure, like school and health care networks, is starting to fade, and municipalities struggle to come up with innovative ways to promote economic development to expand tax revenues and attract new inhabitants. But the region is also an emerging economic hub, marketed and commodified as an Arctic destination of natural wonders and winter experiences. The municipality invests in developing mining industry but also Arctic wild food businesses and fishing tourism, even though the industries are incompatible in values and environmental impact. Locals live alongside the dystopia of closed company villages, a fading welfare state, and the nature and tourism utopia of silence, untouched nature, and superfoods. In this landscape of ecological, political and economic changes and aspirations, the speculation of re-reopening the iron ore mine takes place (Keskitalo et al. 2019). Depending on according to which politics the mining assemblage is organized, the nature gets different values, too. The capitalist mining assemblage foregrounds nature as commodity, while the state mining assemblage underlines the investment in infrastructure, permanent jobs and tax revenue. The nomadic assemblages can be found in some local environmental movements, emphasize the important of nature to the survival for all life in the region, as well as in some local environmental art projects. The locals tread carefully the rhizome of overlapping and contradictory assemblage politics: in the small community, the network of friends, neighbours and relatives covers the mining assemblages, and politics becomes easily personal and embodied.

References

Aatsinki, Ulla. 2009. *Tukkiliikkeestä kommunismiin. Lapin työväenliikkeen radikalisoituminen ennen ja jälkeen 1918*. Tampere: Tampere University Press. https://trepo.tuni.fi/handle/10024/66426.

———. 2017. 'Rajat ja radikalismi : Pohjoinen työväenliike tiellä vallankumoukseen'. In *Työväki kumouksessa*, edited by Sami Suodenjoki and Risto Turunen, 141–73. Työväen historian ja perinteen tutkimuksen seura. Helsinki. https://helda.helsinki.fi/handle/10138/219029.

———. 2019. 'Lapista Amerikkaan: Siirtolaisuuden Taustoja/From Lapland to America : Background to Immigration'. I *Heillä Oli Unelma: Siirtolaisuus Kemijärveltä Amerikkaan/They Had a Dream: Immigration from Kemijärvi to America*, edited by Olli Säynäjäkangas, 8–16. Rovaniemi: OMS Kirjatuotanto.

———. 2020. 'Pohjoisen valkoinen kontrolli ja vapaussodan jatkuvuus työväestön arjessa'. In *Valvottu ja kuritettu työläinen*, edited by Leena Enbom, Pete Pesonen and Sami Suodenjoki. 51–94. Työväen historian ja perinteen tutkimuksen seura. Helsinki. https://helda.helsinki.fi/handle/10138/219029.

Bell, Lindsay A. 2023. *Under Pressure: Diamond Mining and Everyday Life in Northern Canada*. Toronto: University of Toronto Press. https://doi.org/10.3138/9781487548575.

Borah, Prithiraj. 2021. 'Colonial Construction of the "Lazy Native" and Formation of Assam's Tea Industry'. In *Communities, Institutions and Histories of India's Northeast*, edited by Charisma K. Lepcha and Uttam Lal, 65–80. London: Routledge.

Deleuze, Gilles, and Félix Guattari. 1972/1977. *Anti-Oedipus: Capitalism and Schizophrenia*. Translated by Robert Hurley, Mark Seem, and Helen R. Lane. New York: Viving Press.

Evans, Chris, and Göran Rydén. 2007. *Baltic Iron in the Atlantic World in the Eighteenth Century*. Leiden: BRILL. https://doi.org/10.1163/ej.9789004161535.i-360.

Fagervall, Britt-Inger, Esther Salomonsson-Juuso, and Britta Tervaniemi. 2006. *Kengis bruk : världens nordligaste järnbruk*. Thesis. Luleå tekniska universitet. C-uppsats. DiVA, id: diva2:1022528

Foucault, Michel. 1975/1977. *Discipline and Punish: The Birth of the Prison*. Translated by Alan Sheridan. New York: Pantheon Books

Foucault, Michel. 2007. *Security, Territory, Population: Lectures at the Collège de France 1977–1978*. Edited by Michel Senellart. Translated by Graham Burchell. New York: Palgrave Macmillan.

Haikola, Simon, and Jonas Anshelm. 2020. 'Evolutionary Governance in Mining: Boom and Bust in Peripheral Communities in Sweden'. *Land Use Policy* 93 (April): 104056. https://doi.org/10.1016/j.landusepol.2019.104056.

Hast, Sanna. 2021. *Yhteensovittamattomat Luonnonvarat?: Tutkimus Lapin Luonnonvaraistumisesta*. Acta Electronica Universitatis Lapponiensis. Rovaniemi: Lapin yliopisto.

Isaksson, Pekka. 2001. *Kumma kuvajainen: rasismi rotututkimuksessa, rotuteorioiden saamelaiset ja suomalainen fyysinen antropologia*. Pohjoiset historiat 1. Inari: Kustannus Puntsi.

Jokinen, Mikko. 2019. 'Cultural Factors of Environmental Disputes in Lapland'. *Dissertationes Forestales* 2019 (281). Helsinki: Helsingin Yliopisto. https://doi.org/10.14214/df.281.

Junka-Aikio, Laura. 2023. 'Whose Settler Colonial State? Arctic Railway, State Transformation and Settler Self-Indigenization in Northern Finland'. *Postcolonial Studies* 26 (2): 279–301. https://doi.org/10.1080/13688790.2022.2096716.

Keskitalo, E. Carina H., Maria Pettersson, and Sverker Sörlin. 2019. 'Introduction: Understanding Historical Contingencies into the Future: Cases from Northern Europe'. In *The Politics of Arctic Resources*, edited by E. Carina Keskitalo. London: Routledge.

Komu, Teresa. 2019. 'Dreams of Treasures and Dreams of Wilderness – Engaging with the beyond-the-Rational in Extractive Industries in Northern Fennoscandia'. *The Polar Journal* 9 (1): 113–32. https://doi.org/10.1080/2154896X.2019.1618556.

Komu, Teresa. 2020. *Pursuing the Good Life in the North: Examining the Coexistence of Reindeer Herding, Extractive Industries and Nature-Based Tourism in Northern Fennoscandia*. Doctoral Dissertation, University of Oulu.

Kröger, Markus. 2016. 'Spatial Causalities in Resource Rushes: Notes from the Finnish Mining Boom'. *Journal of Agrarian Change* 16 (4): 543–70. https://doi.org/10.1111/joac.12113.

Kuokkanen, Rauna. 2019. 'At the Intersection of Arctic Indigenous Governance and Extractive Industries: A Survey of Three Cases'. *The Extractive Industries and Society* 6 (1): 15–21. https://doi.org/10.1016/j.exis.2018.08.011.

Kuokkanen, Rauna 2022. 'All I See Is White: The Colonial Problem in Finland'. In *Finnishness, Whiteness and Coloniality*, edited by Josephine Hoegaerts, Tuire Liimatainen, Laura Hekanaho, and Elizabeth Peterson, 291–314. Helsinki University Press. https://doi.org/10.33134/HUP-17-12.

Kuoksu, Erik. 2003. *Familjeregister för Pajala socken, Kengis bruk och Torne lappmarks nybyggen fram till 1760: Pajala socken*. Kiruna amatörforskarfören.

Lane, Pia. 2016. 'Standardising Kven: Participation and the Role of Users'. *Sociolinguistica* 30 (1): 105–24. https://doi.org/10.1515/soci-2016-0007.

Mantila, Harri. 2000. 'Lapin Rajat Ja Murrerajat'. *Virittäjä* 104 (2): 188.

Müller, Leos. 2005. 'The Dutch Entrepreneurial Networks and Sweden in the Age of Greatness'. In Trade, Diplomacy and Cultural Exchange: Continuity and Change in the North Sea Area and the Baltic c. 1350–1750, 58–74. Hilversum: Verloren

Mähkä, Rami. 'Suomen sisällissodan vaikutushistoriat'. *Historiallinen aikakauskirja* 116 (2): 189–96.

Naum, Magdalena. 2018. 'The Pursuit of Metals and the Ideology of Improvement in Early Modern Sápmi, Sweden'. *Journal of Social History* 51 (4): 784–807. https://doi.org/10.1093/jsh/shx011.

Nevala, Kirsi. 2015. *Kerjäläisiä ja jauhomattoja—1860-luvun nälkävuodet muistitietoaineistoissa*. Master's Thesis. University of Tampere. https://urn.fi/URN:NBN:fi:uta-201504291367

Niittyranta, Tapio. 2004. *EU:N Vaikutus Kolarin Kunnan Elinkeinopolitiikan Muotoutumisessa, Kolarin Kunnan Seikkailu EU:N Hankeviidakossa*. Master's Thesis, Tampereen Yliopisto.

Nordin, Jonas M. 2015. 'Metals of Metabolism: The Construction of Industrial Space and the Commodification of Early Modern Sápmi'. In *Historical Archaeologies of Capitalism*, edited by Mark P. Leone and Jocelyn E. Knauf, 249–72. Cham: Springer .https://doi.org/10.1007/978-3-319-12760-6_11.

Nordin, Jonas M., 2020. *The Scandinavian Early Modern World: A Global Historical Archaeology*. London: Routledge.

Nordin, Jonas M., and Carl-Gösta Ojala. 2017. 'Copper Worlds: A Historical Archaeology of Abraham and Jakob Momma-Reenstierna and Their Industrial Enterprise in the Torne River Valley, c. 1650-1680'. *Acta Borealia* 34 (2): 103–33. https://doi.org/10.1080/08003831.2017.1397397.

Nordin, Jonas Monié. 2024. 'Copper, Colonialism, and Local Conflicts: The Expansion of Early Modern Global Industrial Economy in Northern Torne River Valley, and Its Local Repercussions.' In *Minorities in Global History: Cultures of Integration and Patterns of Exclusion*, edited by Holger Weiss, 123–40. London: Bloomsbury.

Nurmi, Risto. 2019. 'A Clockwork Porridge. An Archaeological Analysis of Everyday Life in the Early Mining Communities of Swedish Lapland in the Seventeenth Century'. In *The Sound of Silence. Indigenous Perspectives on the historical Archaeology of Colonialism*, edited by Tiina Äikäs and Anna-Kaisa Salmi, 90–118. New York: Berghahn Books. https://doi.org/10.1515/978 1789203301-006.

Nykänen, Tapio, and Aini Linjakumpu. 2020. 'Laestadians in the Public Sphere: Reading the Biggest Christian Revival Movement in Finland'. In *The Challenges of Religious Literacy: The Case of Finland*, edited by Tuula Sakaranaho, Timo Aarrevaara, and Johanna Konttori, 25–38. Cham: Springer International Publishing. https://doi.org/10.1007/978-3-030-475 76-5_3.

Ojala, Carl-Gösta. 2016. 'Svenska Kyrkan Och Samiska Mänskliga Kvarlevor'. In *De Historiska Relationerna Mellan Svenska Kyrkan Och Samerna: En Vetenskaplig Antologi*, edited by Daniel Lindmark and Olle Sundström, 993–1028. Skellefteå: Artos & Norma bokförlag.

Ojala, Carl-Gösta, and Jonas M. Nordin. 2015. 'Mining Sápmi: Colonial Histories, Sámi Archaeology, and the Exploitation of Natural Resources in Northern Sweden'. *Arctic Anthropology* 52 (2): 6–21.

———. 2019. 'Mapping Land and People in the North: Early Modern Colonial Expansion, Exploitation, and Knowledge'. *Scandinavian Studies* 91 (1–2): 98–133. https://doi.org/10.5406/scanstud.91.1-2.0098.

Paulaharju, Samuli. 1923. *Wanhaa Lappia Ja Peräpohjaa*. Helsinki: Kustannusosakeyhtiö Kirja.

Pietikäinen, Sari, and Kori Allan. 2020. 'Jobs for Life?': Mining Temporalities in a Transforming Arctic Periphery'. In *Language, Global Mobilities, Blue-Collar Workers and Blue-Collar Workplaces*, edited by Kellie Gonçalves and Helen Kelly-Holmes, 27–48. London: Routledge. https://doi.org/10.4324/9780429298622-2.

Pietikäinen, Sari, Leena Huss, Sirkka Laihiala-Kankainen, Ulla Aikio-Puoskari, and Pia Lane. 2010. 'Regulating Multilingualism in the North Calotte: The Case of Kven, Meänkieli and Sámi Languages'. *Acta Borealia* 27 (1): 1–23. https://doi.org/10.1080/08003831.2010.486923.

Pietikäinen, Sari, and Helen Kelly-Holmes. 2013. Eds. *Multilingualism and the Periphery*. Oxford: Oxford University Press.

Regnard, Jean-François. *Voyage en Laponie*. 1681/1731. Translated Marja Itkonen-Kaila 2012. Rovaniemi: Väylä-Yhtiöt.

Rydén, Göran. 2002. 'Vallonbruk, vallonsmeder och vallonsmide—en precisering av ett kunskapsläge'. In *Valloner—järnets människor*, 290. Umeå: Umeå Universitet.

Salokorpi, Asko. 2007. *Suomen Rautaruukit*. Lahti: Otava.

Schoderer, Mirja. 2023. 'Overlapping Governmentalities and the Cosmo-Politics of Mongolian Water- and Miningscapes'. *Geoforum* 145 (October): 103830. https://doi.org/10.1016/j.geoforum.2023.103830.

Scott, David. 2005. 'Colonial Governmentality'. In *Anthropologies of Modernity*, 21–49. John Wiley & Sons, Ltd. https://doi.org/10.1002/9780470775875.ch1.

Similä, Jukka, and Mikko Jokinen. 2018. 'Governing Conflicts Between Mining and Tourism in the Arctic'. *Arctic Review on Law and Politics* 9 (June): 148–73. https://doi.org/10.23865/arctic.v9.1068.

Singleton, Theresa A. 2001. 'Slavery and Spatial Dialectics on Cuban Coffee Plantations'. *World Archaeology* 33 (1): 98–114.

Sioh, Maureen. 2010. 'Anxious Enactments: Postcolonial Anxieties and the Performance of Territorialization'. *Environment and Planning D: Society and Space* 28 (3): 467–86. https://doi.org/10.1068/d9408.

Snellman, Hanna. 2005. *The Road Taken: Narratives from Lapland*. Inari: Kustannus-Puntsi. https://doi.org/10.31885/2018.00014.

Sörlin, Sverker, Brigt Dale, Arn Keeling, and Joan Larsen. 2022. 'Patterns of Arctic Extractivism: Past and Present'. In *Resource Extraction and Arctic Communities: The New Extractivist Paradigm*, 35–65. Cambridge: Cambridge University Press. https://doi.org/10.1017/9781009110044.005.

Spangen, M., A.-K. Salmi, T. Aikas, C.-G. Ojala, and J. M. Nordin. 2015. 'Mining Sapmi: Colonial Histories, Sami Archaeology, and the Exploitation of Natural Resources in Northern Sweden'. *Arctic Anthropology* 52 (2): 6–21. https://doi.org/10.3368/aa.52.2.6.

Suopajärvi, Leena. 2015. 'The Right to Mine? Discourse Analysis of Social Impact Assessments of Mining Projects in Finnish Lapland in the 2000s'. *Barents Studies* 1 (3): 36–54. https://urn.fi/URN:NBN:fi:ula-201508031265

Tasala, Markku. 2000. *Metsäkaarti—Kolarin Metsäkaartin Jatkosota Ja Rauha*. Oulu: Pohjoinen.

Tolvanen, Anne, Pasi Eilu, Artti Juutinen, Katja Kangas, Mari Kivinen, Mira Markovaara-Koivisto, Arto Naskali, Veera Salokannel, Seija Tuulentie, and Jukka Similä. 2019. 'Mining in the Arctic Environment—A Review from

Ecological, Socioeconomic and Legal Perspectives'. *Journal of Environmental Management* 233 (March): 832–44. https://doi.org/10.1016/j.jenvman.2018.11.124.

Trevor-Roper, Hugh R. 1980. *History and Imagination*. Oxford: Oxford University Press.

Tsing, Anna. 2016. 'Earth Stalked by Man'. *The Cambridge Journal of Anthropology* 34 (1): 2–16. https://doi.org/10.3167/ca.2016.340102.

Tuominen, Marja. 2003. 'Lapin Sodan Tuhot, Materiaalinen Ja Mentaalinen Jälleenrakennus'. In *Lappi, Maa, Kansat, Kulttuurit*, edited by Ilmo Massa and Hanna Snellman, 102–119. Helsinki: Suomalaisen Kirjallisuuden Seura.

Tuori, Kaius. 2015. 'The Theory and Practice of Indigenous Dispossession in the Late Nineteenth Century: The Saami in the Far North of Europe and the Legal History of Colonialism'. *Comparative Legal History* 3 (1): 152–85. https://doi.org/10.1080/2049677X.2015.1041732.

Valkonen, Sanna, Áile Aikio, Saara Alakorva, and Sigga-Marja Magga. 2022. *The Sámi World*. London: Routledge. https://doi.org/10.4324/9781003025511.

Vilkuna, Kustaa H. J. 1996. *Arkielämää patriarkaalisessa työmiesyhteisössä: rautaruukkilaiset suurvalta-ajan Suomessa*. Historiallisia tutkimuksia 196. Helsinki: SHS.

Yle Lappi. 2021. Regional Television News. Finland. Accessed December 13, 2021.

4

Taste of Commons: Revaluing Arctic Wild Berries

4.1 Berry Fever

Fig. 4.1 Cloudberry, the Lapland gold

S. Pietikäinen, *Cold Rush*, Language and Globalization,
https://doi.org/10.1007/978-3-031-63995-1_4

Every summer, thousands of hopeful berry pickers, armed with mosquito repellent, big buckets, and an appetite for long hikes, venture into Lapland forests, bogs, and marshlands in search of wild Arctic berries. The jewel in the crown of berries is the cloudberry, the first to ripen in late July at the edges of Arctic boggy areas. This rare and valued berry, affectionately known as Lapland Gold, gets its name from the golden hue of the ripe berry and its exceptional nutritional and economic value, which surpasses that of other Arctic wild berries. The nickname also resonates with significant gold reserves and mining operations in the region, including Europe's largest gold mine. The rarity of the cloudberry, coupled with the labour intensive process of foraging, makes it a true delicacy. A luxury berry beloved by chefs, health-conscious consumers, tourists, and locals alike. The cloudberry's distinct, sharp flavour can be found on the menus of high-end restaurants, family celebrations, homemade jams, and in the bucket lists of tourists. The cloudberry has a special meaning to my family: Our oldest child's name is *Hilla*, which is a local Meänkieli language/Torniojoki dialect term for cloudberry (Fig. 4.1).

Cloudberries are a very delicate berry. Summer frosts in June can "bite" the blooming flowers, destroying harvest hopes. Too little, or too much water impacts the size and taste of the berry.

Yet, in the vast Finnish Lapland terrain, there are always perfect local growing conditions for cloudberries, creating pockets of cloudberry fields and a berry hunter's heaven. As the short Arctic summer progresses, speculations about locations and the quantity of cloudberries grow and intensify. The berry fever begins to spread, echoing the salmon fever in the river valley (see Chapter 1). The non-stop coverage of the speculation, price information and weather conditions by local newspapers and radio stations fuels the fever, prompting berry hunters to drive hundreds of kilometres to treasure hunt.

The berry buzz is intense during the few brief summer weeks that mark the cloudberry season. As the summer sun does not set in the Arctic region, berry hunting takes place around the clock. Berry pickers come with different stakes and resources. The locals slip into the forest and head towards their designated berry patches, the location of which is kept a family secret, in the mornings, when the wind helps keep the mosquitos at bay. Only the perfect berries—glimmering orange, dry, and perfectly ripened under the Arctic sun—find their way into the buckets. The berries are frozen immediately upon returning home to be enjoyed during the winter. Other berries are left to ripen and to be gathered on subsequent mornings. The hastily parked cars along the roadside of the near promising-looking bog lands belong to berry hunters from elsewhere in Finland who have travelled to Lapland's berry fields to try their luck. Some stay for a day, some a few days longer, camping in the forest and diligently picking the berries around the clock to fill up their winter reserves.

As much as I enjoy eating wild Arctic berries, I am not a big fan of foraging for them. I am happy to pick mushrooms on the cool autumn days, but picking berries is hard work on a warm summer day in the company of an army of mosquitos. It involves walking long distances through marshy fields (if looking for cloudberries) or in dry pine forests (for blueberries), while carrying baskets, a packed lunch, and mosquito repellents. Until recently, I have not had any compelling reason to go for serious berry picking as I have been supplied with plenty of berries by a network of aunts, my mother-in-law, and my mother. However, with the passing of that generation or their increasing frailty precluding berry picking, I found myself in search of berry supplies, and it has been rather hopeless. The best, freshly picked berries are nowhere to buy: the berries sold at the grocery shops have travelled a long distance and are picked in haste by the commercial pickers. They do not taste the same as the plump, ripe berries picked by the hands of aunties and mothers.

Occasionally, I purchase some berries from a retired local teacher and an industrious berry picker, who sells them from his car trunk by one of the main roads in Lapland. But these berries are typically already frozen, and I prefer fresh berries. Over the past few years, through my local network of friends and relatives, I have gradually been upgraded within the berry network to a permanent berry waiting list: this means that if local pickers or their picker friends happen to want to sell some portion of their berry harvest, I might be able to purchase them. In the queue, there are no discussions about the quantities or costs of the berries, only the types of berries (cloudberries, blueberries). We all recognize that being accepted in the queue is a privilege that we value highly.

If I am lucky, I may receive a call on a summer evening telling me how much and what type of berries I will get. Some years, I have gotten 10 kg of fresh cloudberries, sometimes only a few kilos, and on some summers, none at all. Blueberries are easier to source and I usually get between 10–20 kilos.

During the last few summers that my mother-in-law spent in Lapland, in the cottage built on the lands of her childhood home, she talked to me about the family berry patches nearby. I wish I had paid closer attention. Nowadays, when spending my summers at the cottage, I have made a few short expeditions in the nearby areas to identify patches available for me to pick berries. I have picked some blueberries near the cottage but have not found a promising patch for cloudberries. I am still in the process of becoming a berry picker.

In this chapter, I examine the revaluing processes of Arctic wild berries from commons to commodities with the help of critical assemblage analysis (see Chapter 2, also Deleuze and Guattari 1984/1987). To create different values for the berries, they are fitted in different yet overlapping assemblages organized by different logics. These are berry commons assemblage, commercial berry assemblage, and state berry assemblage. Each entails dynamics of re/deterritorialization and recoding, reconfiguring the relations between berries, bodies, and land.

Arctic wild berries have a significant role in Arctic life, constituting a vital component of Arctic self-sufficiency (Joks 2022; Teetl'it Gwich'in Renewable Resources Council et al. 2006). Whether preserved as jams or juice, dried or frozen, the berries are valued as a much-needed source of nutrition in a harsh Arctic climate with limited possibilities to harvest or grow fruits and vegetables (Nilsson 2022). The value of wild berries is captured in the slogan "one cloudberry contains more vitamin C than an orange", which codes berries as nutritionally superior to fruits typically imported from the South. Apart from their nutritional value, Arctic wild berries have been used for exchange and trade, embedded in indigenous governance of resources and relationships (Joks 2022; Länsman 2004; Teetl'it Gwich'in Renewable Resources Council et al. 2006). Now, they are being increasingly revalued as natural resources used for Arctic economic development for global markets.

In Finland, some 42 million kilos of berries are harvested annually for domestic consumption, while about 22 million kilos are picked for the berry industry. The export of berry products has increased rapidly due to new preservation technologies as well as the expanding market for new berry products, such as cosmetics, oils, and teas. The turnover of the natural product sector was 780 million euros in 2020 (Ministry of Agriculture and Forestry for Finland 2024). Despite these economic prospects, some 90% of the wild berries are left unpicked in the Finnish forests every year.

The speculation around the Arctic berry rush is intensifying, spurred on by expanding global markets for pristine berries and the economic potential of these seemingly "free" commons abundantly available in the forests. This surge has given a boost to numerous new small companies and development projects seeking to capitalize on the berry rush. Simultaneously, conflicts surrounding the berries have also intensified, often tied to their status as commons (Hedberg and Olofsson 2022; Huhmarniemi 2012). In Finland, the right to forage for berries, mushrooms, and herbs, as well as access to forests, is granted freely under a traditional common law known as *jokaisen oikeus* or "everyone's rights" in English (Tuunanen et al. 2012). In practice, this means that everyone—local, tourist, visitor—is allowed to pick up berries and mushrooms for their own use anywhere in the vast forest areas, provided that they do not disturb local inhabitants or cause any harm to nature.

However, simultaneously with this legal berry cartography, there also exists an established social berry cartographies and contracts rooted in the indigenous practices of resource governance (Joks 2022; Länsman 2004, Nilsson 2022). Consequently, Arctic forests are mapped out and territorialised into distinct berry patches, each belonging by usage to certain families and regulated by enduring social norms. Arctic berry territories are not neutral terrains for foraging; rather, they are governed by social networks and imbued with political and economic interests. While wild berries are legally deemed "free" for anyone to forage, in practice, this freedom is limited by historical usage and social relations. Foraging Arctic wild berries is intricately woven into a complex rhizome of social norms, relations, economic potentials, and emerging commodification practices.

Commercial wild berry picking is in sharp contrast to this. Over the past 20 years, the Finnish Arctic berry patches have turned into migration fields, with imported commercial pickers, originally from Ukraine, Russia and Latvia, and more recently from Thailand, working for Finnish berry companies for some weeks during the berry high season. Currently, migrant berry pickers forage 92% of all wild berries for commercial usage in Finland (Marsi 2019). Pick-up cars and minivans with big Finnish berry company slogans, carrying groups of imported Thai migrant commercial berry pickers, circulate in the small Lapland villages and roads. Distinctive with their yellow vests, obligatory for the imported pickers to wear for their safety and visibility in case they get lost in the forest, the pickers move in groups equipped with GPS, maps, and commercial berry picking equipment, namely a particular kind of brush. Berry companies rent old abandoned village schools, camping, or caravan parks that are no longer in commercial use to house migrant pickers. In the peripheral Arctic villages, far from the tourism resorts, the rare encounters between the migrant pickers and locals happen in the few remaining village shops and gasoline stations. Material signs of the new customers include new kinds of food ingredients in the shops and a few signs in English saying "no currency exchange".

The unevenly unfolding Arctic berry economy treats the locals and migrant berry pickers differently. Besides the cultural and geographical know-how for Arctic foraging, the locals have unchallenged access to the berry patches. Furthermore, the state encourages small-scale entrepreneurship related to berry foraging by exempting the Finnish pickers from taxation. This opens up a possibility to earn extra money without the high rate of Finnish taxation—a possibility that is hoped to attract students, the unemployed, and retired people especially. However, the economic prospect is not the main attraction for berry picking; rather, it is part of domestic practices and free time activities. This is also reflected in the typical Finnish bodies engaging with foraging: senior female bodies, the aunts, mothers, and grandmothers. For migrant pickers, foraging is hard labour, and the bodies of the migrant berry pickers are typically male and from the rural countryside, accustomed to hard physical labour (Hedberg 2013; Hedberg and Olofsson 2022; Olofsson et al. 2023). To make enough profit after paying travel and accommodation costs, the migrant pickers must pick a significant number of kilos daily, regardless of the weather or berry season. They drive along the Lapland roads in search of good berry patches, aided by previous years' experiences, maps, and advice from the berry company. The foraging is done in big groups, using all available equipment. The working conditions for migrant pickers are hard, even in the best circumstances, and some of the Finnish berry companies have been accused of and convicted for unethical and criminal practices, including human trafficking (Ylikoski 2024).

The bodies that pick the berries also create different values for the berries. The discursive valuation of the labour by the different types of pickers affects the value of the berries and positions the consumers differently (Radovic-Fanta 2021). Hand-picked is a new category emerging from the need to differentiate berries picked by locals and can be argued to be part of the green, local economy and sustainable practices, thus justifying the higher price. Hand-picked is an attribute used in commercial discourses globally to index rarity and luxury (see, e.g. Li and Leonas 2019). Unmarked berries usually index commercial picking, also visible in the number of leaves among the berries, resulting from the equipment used in commercial foraging, explaining the lower price. Migrant pickers connect the berries at the minimum with the commercial terrain and potentially with unsustainable and unethical practices, while the local pickers fit with the idea of the local economy and sustainable practices. Commercial picking with migrant labour reconfigures both the relations around the berry and the value of the berry, creating a new distinction between commercial berries and hand-picked berries. Migrant labour and commodification of wild arctic berries is a process of uncommoning the commons.

Arctic wild berries are simultaneously fitted into three different yet overlapping assemblages: commons, commercial, and work assemblages, organized by different politics, namely territory, capitalism, and state respectively. The revaluing of berries is not simple or straightforward, and it involves a series of speculations and experiments, some successful, but not all of them. Revaluing berries also raises complex questions related to access, sustainable use of commons, fair employment practices, and sustainability of the green economy. The reterritorialization of the berries from ecology and local subsistence to economy entails the reconfiguration of material and discursive elements as well as social relations governing natural resources. The new meanings of the berries are created by recoding the berries as superberries and foraging as green, sustainable work. As we shall see, the first one works, while the latter fails. In the midst of these changes, the meaning of berries as commons endures, and locals push back.

Berries make things happen: they reshape social relations, transform the subsistence economy, and contribute to the region's economic development. They belong to the indigenous governance system of natural resources (Joks 2022; Nilsson 2022; Teetl'it Gwich'in Renewable Resources Council et al. 2006) and they deserve scrutiny in and of themselves. In this chapter, I examine different values, practices, and experiences around Arctic wild berries in the context of the Cold Rush. The overlapping, intertwined cartographies of Arctic Berry patches connect them as part of global health markets, migrant labour, local economic reinventions, and historical commoning practices. The new attempts to revalue the berries from commons to commodities simultaneously create tensions and possibilities, and I want to examine how people navigate these complex and changing berry patches, moments when people, berries, and their social, cultural, environmental, and economic valuation come together as assemblages, some functioning better than others.

I examine berry assemblages and how they work (or not) with the help of two local berry pickers, Leena and Heli. Both are female local pickers with different trajectories and access to the berries. Leena, a senior picker, has been foraging the berries for decades from her family berry patches. She is considered one of the most knowledgeable and successful berry pickers in her village. She has always done berry foraging, but especially since she retired from administrative work in the public sector, she has been able to devote more time to foraging. Now, aged over eighty, the berry trips have become shorter and rarer, but they still belong to her summer practices. Through marriage, Leena and I belong to the same family and village networks, which allows me to reflect on my own foraging practices and access to berry patches, and the relationships between gifts and exchange. After my mother-in-law's passing, Leena has become the person to whom I turn to ask for information about berries and how to preserve them, as well as for stories about berry patches and the old ways of foraging.

Heli worked for a local berry economic development project funded by the EU and the local municipality. The three-year project aimed to create a blueprint for a wild berry value chain that would create local jobs and was connected to broader green economy development aspirations in the area. Heli is from the region, and after living in several different countries and doing a wide range of jobs, she wanted to return to the Arctic and re-rooted herself in the area by building a house and getting a local job. She worked first in the tourism business and then got a job as a project coordinator in the berry project. I followed Heli in her work throughout the three-year project. This involved, among other things, meetings with local stakeholders, training potential pickers, and project development from the initial idea to the final report. We discussed the various steps and turns related to the project and the possibilities and constraints of creating berry picking as a sustainable work opportunity for local people.

These two berry pickers are part of overlapping yet distinctive berry assemblages with different elements, organizations of relationships, and functions. Leena functions in a more traditional, domestic subsistence assemblage organized around the shared idea of commons and reciprocal gifts. Heli is part of the economically oriented assemblages, organized by a politics of state with interest in creating jobs and tax revenue, but impacted by a capitalist logic of profit. The assemblages include Arctic Wild berries, bodies that do the picking, and the act of picking, but they also have incorporated different elements to gain their full capacity. In the following, I will discuss each assemblage and how the value of berries gets reconfigured.

4.2 Berry Gifts and Thefts: Rhizome of Governing Commons

Fig. 4.2 Arctic wild blueberry

Leena is one of the best berry pickers in her village when evaluated on the quantity of quality berries she can find and pick every summer, regardless of the changing, unpredictable growing conditions during the short Arctic summer. This is not a coincidence or sheer luck but the result of enduring work and in-depth knowledge of the terrain, berries, and weather of her berry patch area. I have watched her take long hikes in the forests and boglands near and further away from the village to check her berry patches, whose precise location is known only to her and her husband. She starts her hikes already in June when the berries begin to bloom. I have been listening to her berry predictions for the summer over cups of coffee, a necessary element of our regular visits to each other. When the berry-picking season starts in late July, I see Leena coming and going into and about the woods, her house, village road, and having a bucket full of berries at the end of each picking trip (Fig. 4.2). The evenings are busy for her as she cleans and freezes the berries right away to preserve them at their best for the winter. Every once in a while, she gives me tastings, freshly baked berry pie, a packet of frozen berries, or a jar of jam. I try to reciprocate by offering her a lift to the nearest town, bringing her a few groceries from the market, or helping out with heavy-lifting home chores.

One exceptionally hot summer day during the berry season, Leena, well in her 70s at the time, told me that her health had taken a bit of a setback, and she was somewhat apprehensive about going deep into the forest to pick berries. I interpreted this as an indirect way of asking me to accompany her on a foraging trip, and I volunteered. To my surprise, she adamantly refused my help, bushing me off lightly but firmly with a laugh. With embarrassment, I realized that I had overstepped a mark. Despite our long friendship and the fact that we were married into the same local family network, the berry patches are shared only with the inner circle of family members. Even though her children or grandchildren were not there to accompany her for her foraging tips, she still wanted to reserve her berry patches only for them.

To restore our relationship, she gave me three berry gifts. Later that evening, she brought me a small box of fresh, golden cloudberries, still smelling of the forest. Later that week, she offered to go with me to forage berries further away from the village. This offer did not include sharing her family's berry patch but instead sharing her knowledge about the area and territories free for berry picking. And, later that summer, when I was already back in my Southern urban home, she called me and asked if I wanted to buy very good quality cloudberries picked by her friends, a retired couple who wanted to earn a little extra. This was a precious gift. It is extremely hard to find locals who want to sell their berries, and as there are always more people who want to buy directly from locals, to get a chance to buy berries is a gift. I eagerly said yes and Leena told me she would buy and freeze them for me, a gift with wrapping. Even today, when Leena is over eighty, and her health prevents her from going for long berry-picking trips, she is not inviting me to come along, and I dare not offer again. I have realized that it will never be my place to know her berry patches.

As mentioned earlier, Arctic berries are not free resources available for anyone to pick from anywhere. Despite their legal status as commons, berry foraging is governed by various social systems and networks related to families, belonging, and land (Nilsson 2022). In the Finnish Arctic, a historical governing system of commons was an indigenous Sámi system called *Siida* in the Northern Sámi language (Lehtola 2012). *Siidas* were autonomous indigenous communities with common usufruct territories for their livelihoods, including reindeer herding, fishing, hunting, and foraging (Kulonen et al. 2005: 392, see also Chapter 2). This indigenous governance system was gradually overridden by settler colonialization beginning in the seventeenth century (Spangen et al. 2015), but we still have echoes of it in present-day foraging practices (Joks 2022).

As the settler colonialization expanded, it brought a new kind of competition over resources, including berries. Under these changing conditions, a reciprocal gift exchange system between the indigenous Sámi and Finnish settlers gradually developed to ensure recognition of social norms related to land use and to manage the redistribution of natural resources, especially foraging and fishing waters. The system is called *verdde* in Northern Sámi and *väärti* in Finnish (Länsman 2004), Finnish *väärti* being a loan word from the Northern Sámi word *verdde,* which means a friend and is also used to refer to these reciprocal relations (Länsman 2004). Originally, the *verdde/väärti*-system referred to an exchange system between two economies: the Finnish farming economy and the indigenous subsistence economy. A person categorized as *verdde/väärti* was someone between a friend and a guest in terms of closeness, while the category's long-term temporality turns it into a relationship. Later, *verdde/väärti*-relationships expanded to refer to exchange between the indigenous Sámi people and southern Finnish visitors, and even between local inhabitants and southern visitors. (Länsman 2004: 43–48, Lehtola 2012: 37–53, 238–242.)

In the exchange economy governed by the *verdde/väärti*-system, berries gained the value of gifts. In a productive, reciprocal *verdde/väärti*-relationship, southern visitors would bring goods that were difficult to get in the North, such as spices, coffee, and spirits (Länsman 2004). In return, the Sámi host would guide visitors to good berry patches and fishing waters. Gifts are complex social practices governed by respect, balance, and obligations (Mauss 1954; see also Fabinyi 2013; Osteen 2002; Yan 2012), and they are used to build trust, hierarchy, and solidarity (Nilsson 2022; Peebles 2015). The *verdde/väärti*-system reconfigured the relationship between the Sámi and newcomers.

With Arctic berry gifts, recognition and reciprocity are essential. Recognition relates to understanding and respecting the social norms and boundaries regulating the use of berry patches as well as appreciating the knowledge and labour that goes into foraging them (Joks 2022). Varying weather conditions impact when and where the berries grow, and finding the best patches for each season requires ongoing monitoring of weather as well as long hikes in the forests and wetlands to check on the blooming, ripening, and picking conditions. Reciprocity validates the match between the gift and counter-gift: if the value difference between gifts is too large, it disturbs the exchange and the social relations in which it is embedded. Gifts that are too valuable put the receiver in debt, while gifts that lack sufficient value indicate a lack of respect and recognition and can cause offense. Both cases harm the relationship between the parties. Consequently, navigating the gift exchange requires a sensitive understanding of subtle nuances and boundaries.

While the *verdde/väärti*-system has become less recognized today (Länsman 2004: 87), it continues to resonate and regulate the expectations, norms, and relations around berry-picking. Over time, the idea that certain places belong to certain families by social contract has expanded beyond indigenous contexts to berry-picking throughout the Finnish Arctic region. Regardless of their background, locals distinguish between dedicated *marjapaikka,* berry patches of land where a particular family has a recognized privilege to pick berries and between the forest and uncharted terrain that is available for anyone to forage. The Arctic territory is charted by an unmarked map of berry patches belonging to certain families, a relationship established through usage, not legal ownership (Joks 2022: 154).

I now understand why my mother-in-law tried to involve me in berry picking in the Arctic: she was trying to pass on the knowledge of her berry patches to me, but at the time, I was a young, busy mother of three kids and I did not recognize the value of this invitation. Foraging is an embodied practice in nature that has been charted by a cultural cartography of family berry patches and biological and geographical knowledge and experience of the places where the berries are most likely to be found in different summers with different growing conditions (Joks 2022; Siitonen et al. 2021). Joks (2022: 153) explains that in Northern Sámi, cloudberry picking areas are called *luomemeahcci*, which can be translated as a cloudberry taskscape, emphasizing the intertwined relationship between activity and land (see also Trusler and Johnson 2008). I can still find berries in my childhood forests, where I foraged with my mother for my first 20 years. And I am still not good at finding berries in the Arctic region, despite over 30 years of spending summers there.

Failure to recognize the social governance of berry foraging or breaking the unspoken practices turns berry-picking into a disrespectful activity, even theft, and changes productive social relationships into violations and abuses of trust (Länsman 2004; Joks 2022). Locals have developed different strategies to navigate changes in berry patches and to mitigate possible competition or conflicts. One is silence. To protect their berry patches and social relationships in their networks, locals tend to be silent about their berry patches as well as cautious and respectful of others. Silence also includes visual silence: a friend of mine, an enthusiastic berry picker and Instagrammer, never posts pictures of her foraging trips in order to conceal geographical information about her berry patches. Like Leena, locals do not disclose the locations of their berry patches to anyone outside the family, but, rather, if asked, give a general description related to the area's geography, for example, a specific lake upstream of the river valley, etc. Insisting on information about a specific location is also a violation of social rules and is considered disrespectful, as I was reminded by Leena's gentle refusal.

Commercial berry picking is in sharp contrast to the social governance of berry commons and the relations around them. It requires a reconfiguration of the berry assemblage and the creation of a commercial berry assemblage that functions along capitalist logic. At the same time, the berry commons assemblage endures and occupies the same terrains. The two assemblages overlap and co-exist, but the dividing difference is the view on land: the commercial berry assemblage views the berry terrains as frontiers while the berry commons assemblage codes them as berry patches, already occupied, marked by historical usage. These berry patches belong to certain families, and others are expected to respect this. This social usage and ownership govern all berry patches near villages and houses. Although for outsiders, the terrain might look empty and unoccupied, the local foraging map has traces in the territory, and those with knowledge can read them. However, the printed maps by the state forest board, available for commercial pickers, represent the area as a vast common area for anyone to come and pick. Similar to the extractive mining industry (see Chapter 3), commercial picking practices abide by a colonial view of empty land as a resource frontier available to anyone for extraction.

The berry patches near villages and houses become a site of tension when different economic rationales overlap the same territory (Boulanger-Lapointe et al. 2019). Even though over 90% of the berries are left unpicked every year in the vast Finnish Lapland forests, tensions between local and commercial berry pickers arise when the two foraging practices overlap on the same terrain. Many local pickers tell me "berry war" stories when I ask about commercial pickers: berries are being picked when not yet ripe, the ground is stomped over, and picking happens in socially forbidden places, such as too close to houses and villages, or in the family berry patches. In the Arctic, commercial berry picking practices disregard the commonage practices and the entire set of relations, meaning, and practices around foraging, which govern what commons are being defined as resources, which communities of people are seen as holding rights to them as well as how regulations are developed and what modes of enforcements are set in place (Natorzky 2013: 123).

At the local Arctic berry patches, tensions between local berry pickers and imported migrant berry workers stem, on the one hand, from a lack of awareness of the social norms governing berry picking and, on the other, the precarity of imported workers, who must pick as many berries as possible in the shortest time possible, which often includes picking underripe berries and using machines (Eriksson et al. 2019; Lacuna-Richman 2021; Rantanen and Valkonen 2011). Locals feel violated after finding their berry patches emptied or encountering migrant pickers in the terrain the locals understand as their backyard (Länsman 2004). In addition, locals often view commercial picking practices as unsustainable. Heli explains how she can always tell where a group of commercial berry pickers has foraged because of the clear trail they left behind: the machine picking leaves the berry bushes empty (while in hand-picking, the unripe berries are left to ripen), the leaves are bruised, and the ground stomped on.

The spatio-temporal and affective difference between berry commons assemblage and capitalist berry assemblage becomes evident: the local pickers return to the same berry patches, territorialized as "their own berry patch" year after year, while the commercial pickers, if returning at all to Finland for another summer, will work across "Arctic frontiers", in regions where the berry company predicts that the best berry crops are most likely in a particular summer. Also, the temporality between commercial and local picking differs: the capitalist logic that forces the migrant pickers to pick as much as possible as quickly as possible has a ripple effect on the foraging rhythm of the locals: they, too, feel a need to be alerted to a new berry foraging temporality, in fear that the commercial pickers will "empty the berry patches." A local berry picker described how the summer of 2020, when the global pandemic prevented the migrant pickers from arriving, was like a reminder of the old times before commercial pickers: there was no rush for the locals to pick as much as possible as soon as possible. Rather, the locals had the time to pick at a leisurely pace, according to their own pace, without a worry that the commercial pickers might empty the bushes before them.

The berry war stories make visible the berry politics and their value as commons (Feeny et al. 1990: 5; Narotzky 2013: 124). Different berry pickers functioning in different assemblages have different stakes. Commercial pickers test the community ties around berry foraging and bring the global temporality of commodification and profit to berry patches. At the same time, migrant berry pickers connect Arctic commercial foraging to the critical and ethical questions related to the use of precarious migrant workers in farming and foraging globally. As elsewhere, also in Finland, the use of migrant berry pickers has been critically discussed in terms of unethical practices, inequalities, and human trafficking. In the Finnish Arctic, commercial berry companies started to import commercial pickers in the mid-1990s as a response to the lack of local berry pickers and their reluctance to sell their berries in a situation when the global markets were growing (Huhmarniemi 2012). In Finland, the biggest number of pickers, around 4000 annually, come from Thailand (Turtiainen and Rantanen 2020). The idea for bringing in berry pickers from Thailand was borrowed from the neighbouring Swedish Lapland, where migrant Thai pickers have been used since the early 1990s (Hedberg and Olofsson 2022; Olofsson 2024). Like in Sweden, in Finland, the importation of the pickers from Thailand was a result of the combination of informal networks, translocal brokers, and investment of berry companies. In Finland, the pickers come with a tourist visa, allowing them to do foraging for a maximum of three months. The difference between a working visa and a tourist visa is vital here: the latter places the pickers in a position similar to a kind of self-employed entrepreneur, outside of Finnish labour laws or responsibilities for the state or municipality, creating a kind of berry Wild West with colonial echoes (see also Hedberg and Olofsson 2022: 41). This has led to several instances of misuse of the migrant labour, with subsequent lawsuits and eventually, in 2021, a new act regarding using migrant labour in foraging, established more regulation for the berry companies.

The commercial use of Arctic wild berries connects them with other agricultural industries in rural areas around the globe that rely on seasonal migrant workers (Hedberg 2021), as well as to the well-known problems related to inequality and precarity of migrant workers. Even under the best circumstances, the profit from foraging is scant. Similar to other migrant workers elsewhere taking part in berry, fruit, and mushroom harvesting for global consumption (see e.g., Fabinyi 2013; Veloz 2020), the goal for commercial pickers of Arctic berries is to pick enough berries over the few weeks working in Finland to gain annual income to bring back home. This means some 100 kilos of berries per day. However, the potential income from foraging is also very unpredictable due to the changing Arctic weather conditions and various expenses (housing, traveling, visa) taken out of the berry income. The work is hard, and the living conditions in the camp site or abandoned old school buildings are harsh, creating an on-going debate about whether the use of berry pickers is a form of work migration or exploitation (Rantanen and Valkonen 2011).

Commercial berry foraging deterritorializes berries from the local social governance system and relations and reterritorizes them into the colonial and national imaginary of the Arctic land as the no man's land, the last frontier, free for extraction. Perceiving the berry lands in these resource extraction terms means that different rules apply: berries are free for anyone to extract. In assemblage terms, this can be seen as the emergence of those different assemblages: capitalist berry assemblage and commons berry assemblage, both having berries as one of the central elements but different discourses, logics, and politics. The latest version of the emergent berry assemblages is the superberry assemblage, which brings together wild arctic berries, discourses of sustainability and wellbeing, and the expanding health markets.

4.3 Becoming Superberries: Plotting Against and With Capitalism

Fig. 4.3 Arctic midnight sun

On an August afternoon, at the height of the blueberry season in the Arctic, I watched Heli, a municipality berry project worker, explain to a group of local berry enthusiasts the difference between a super-blueberry and an ordinary blueberry. In her talk, Heli made the distinction based on the Arctic environment and sustainable foraging practices. The first factor roots the berries in the exceptional growing conditions related to the pure nature and the amount of sunlight during the Arctic summer when the sun does not set at all, known as white nights or the midnight sun in English (Fig. 4.3). The second factor attempts to move commercial foraging from the ethically problematic industrial picking using migrant pickers to the communal practices of local pickers with connections to social sustainability, decent salaries, and community development within a wide frame of the green economy. The superberries, by nature, would be sustainably and carefully hand-picked and quickly frozen to preserve their quality, creating superberries through super-foraging practices. However, revaluing the berries into superberries was more successful than revaluing commercial foraging as sustainable work. The project succeeded in recoding mundane blueberries into superberries and reterritorializing the berries into the global terrain of health and wellbeing, but it was less successful in converting domestic berry foraging into sustainable local work. The local berry pickers refused the invitation to deterritorialize their foraging practices from the terrain of domestic subsistence and leisure to be reterrotorilized in the terrain of paid work under capitalist berry industry conditions. This refusal resonates with the locals' refusal to abide by the colonialist and capitalist order in the Kengis mine centuries ago (see Chapter 3). These echoes show how capitalist logic and resistance to it endure and re-emerge.

The impetus for envisioning the commodification of Arctic berries for global markets is one example of various economic reinvention strategies of the Arctic municipalities. Projects like the berry project are widely used structures to innovate and govern economic development attempts. With ample funding opportunities from the EU and national funding schemes for the economic development of peripheries, temporary projects have become a permanent element of the Arctic economic infrastructure, creating a particular kind of project economy that resonates with some critical aspects of neoliberalism in the public sector.

This projectification, understood as a temporary, task-focused organizational arrangement of the economy, changes local economic arrangements and is often problematic as they offer, as Fred (2018) puts it, "short-term solutions for long-term problems." With their particular, often rather isolated goals (like the commodification of wild berries), these projects do not engage with the thinking on larger processes, resources, and conditions that together would create more enduring possibilities for economic development in the region. This has led to fragmented, typically overlapping, and often unsustainable goals and solutions offered by the various projects. For example, during the Cold Rush project, the Kolari municipality managed several projects in addition to the berry project, including a fishing tourism project, a revitalizing peripheral villages project, and an Arctic mobile job project. Temporary projects come with low political and economic risk to the municipality: political consensus is not needed regarding investments or a shared long-term vision of future directions (Fred 2018). Rather, the projects are framed as a machine to extract funding from national and supranational bodies to compensate for the lack of nation-state subsidies. Locally, projects create both opportunities and tensions: at their best, projects can seed ideas, resources, and synergy, but, typically, they are too short-lived, too narrow, and too disconnected from other conditions to have any real impact on the economic development of the regions (Hodgson et al. 2019; Jensen et al. 2016).

The Arctic Berry project was operated within this complex terrain, plotting against and with capitalist logic and aligning with sustainability and green economy values. The project was built on speculation about the growth of foraging markets: the goal was no less than designing a blueprint for wild food infrastructure to extract wild superfoods from Arctic forests to global consumers. This is not without examples from the neighbouring Sweden, where most of the blueberries are exported to Chinese markets (Eriksson et al. 2019). With a life cycle of three years, the aim was to create a berry route "from berry tussock to international consumer." Heli, the project worker, explains that it is not only a journey of the berry, but certain ethical values of sustainability are also carved into the package, so lots of other things move as well.

The project was designed to have three successive cycles: incubator, accelerator, and cluster. These neoliberal buzzwords code the berry project as an Arctic example of global economic development. However, the project remained mainly aspirational. The talk in the August afternoon belonged to the incubator phase, which had two specific goals: first, to create a value chain that would increase the value of Arctic wild berries so that it would become possible to pay berry pickers a decent salary, and second, to recode berry-picking as paid work through various professionalization practices and with a decent wage. On paper, the plan was to create a value chain with four activities, each adding value to berry work: creating sustainable work for locals, professionalizing berry-picking, adopting sustainable picking practices, and creating a berry route from the Arctic to the Asian markets. The project aimed to create a blueprint for this value chain and test it in practice.

To produce a conversion from a blueberry into a superberry, Heli assemblages together the Arctic growing material conditions, the discourse of superfood, and the affects related to health and wellbeing. The transformation of wild Arctic berries into superberries requires assembling key material, discursive, and affective dimensions for "super" quality as well as global circulation. This involved the de- and reterritorialization of key elements to produce a qualitative change in the value of Arctic berries, making them mobile for global markets.

The materiality of the superberries draws on the idea of the Arctic as the last frontier that naturally produces pure, clean, and safe food. Moreover, the exceptional Arctic growing conditions—around-the-clock sunlight for three summer months—make the wild Arctic berries exceptional in terms of their health potential. The growing conditions lead to the nutritional density which warrants super-qualities for the Arctic berries. To support this claim, Heli presents in her talk recent research on the nutrition densities of Arctic wild berries, creating a link between scientific knowledge, biological environment, and health concerns. They are a superfood by nature, as the marketing slogan goes. Superberries become spatially fixed into pure Arctic nature, attached to the cleanest air and the purest water in Europe. In this spatial fix (Harvey 1981), the local environment is tailored to enhance the value of berry products and to create a marketable difference (Olofsson 2024).

This geographical structuring of the berries also involves discursive structuring: the meaning of wild berries must change to expand their profit potential. Just as the berry industries must adapt to new material conditions like a warming climate and the lack of labour, so too it must reorient its discourse strategies to fit new cultures of food and sustainability. Discourses are, like material infrastructure, significant investments that fix particular values to the berries to make them mobile across fields of consumption and food practices. This simultaneity of material and discursive restructuring is an inherent quality of spatial fixing (Reisman 2019).

Heli incorporates the superfood discourse in the project as part of recoding foraging as paid work. The designation superfood reflects the belief that a single ingredient can possess an exceptional level of nutrition, which is important for well-being and health promotion. Superfoods are celebrated for their purported extraordinary nutritional and/or medicinal value, natural qualities associated with an exotic or pristine growing environment, or their traditional and indigenous use (Loyer 2016; Veloz 2020). In practice, superfood is more of a discourse than a description of a material substance (Sikka 2019), but its power draws on the interaction between material and discursive aspects of the food (cf. Cavanaugh 2007; Shankar and Cavanaugh 2012). Reisman (2019: 340) describes superfood as a culinary meme that has gained powerful momentum through wellness trends.

The moment of sorting the wild berries into the category of superberry is the moment of revaluation: the berries become decoded as commons and recoded as commodities. In physical terms the berries, of course, remain the same, but through this new superfood categorization, their value gets sorted (Dahlin and Fredriksson 2017; Tsing 2015). Arctic wild berries become coded as superberries as part of the region's economic reinvention, fuelled by two trends that have accelerated the expansion of markets for super, local, and ecological food products: wellness and sustainability.

The berries and their growing conditions are coded with superlatives and exceptional qualities. The prefix *super* supplements the Arctic wild berries with qualities of vitality, high spirits, and a glow of overall wellness. Super is a powerful category for Arctic wild berry ingredients and products. Together with other prefixes, such as multi-, trans-, de-, and extra-, super is part of linguistic patterns used to enhance, underline, and expand the quality of an object, approach, or perspective beyond its typical terrain (Reyes 2014). In the case of Arctic berries, super transports them from the realm of a mundane, typical ingredient into the terrain of well-being, vitality, health, and beauty. In this process of de- and reterritorialization, the discourse of Arctic superberries gains its power from superlatives: these berries grow in the purest air in the world, under the never-setting sun, and contain more vitamins than any other berries. The power of nature is supported by visual tropes typically used with superberries: Northern lights, white nights, and pure water—all great wonders of nature, now packed into and delivered by the berries.

The Arctic superberry discourse constructs a causal relationship between the growing conditions and their effects on the consumer. Harnessing the health halo, superberry discourse makes consumers rethink berries beyond their traditional role in the subsistence economy. The health message speaks to concerns related to stress, hectic lifestyles, and improving one's quality of life. Thus, superberry status extends beyond the health-promoting qualities of the berries. It also fosters a consumer identity in which a strategy for well-being, even super powers are needed to navigate the demands of work and personal life. Underlying superfood discourse is the powerful view of eating as a technology of the self (Foucault 2014), through which individuals govern their bodies, well-being, and performance. Superfood discourse resonates with a neoliberal entrepreneurial self, driven by self-improvement (Gershon 2018; Sikka 2019). The connections between pure Arctic nature, health, vitality, and beauty reshape the expectation of wellness from sickness prevention to hyper-productivity. Berries become a vital ingredient for a lifestyle that maximizes output. The superberry discourse turns wild Arctic berries' nutrient density into vitality, which can be marketed and exported to consumers globally.

4.4 Green Foraging: Seeking Right Bodies for Sustainable Berry Picking

Fig. 4.4 Purchased cloudberries from a local picker

The second goal of the municipality berry project, to reterritorialize commercial berry picking into local job opportunities with sustainable foraging practices, turned out to be an impossible assemblage to make function successfully (Fig. 4.4). Combining sustainable berry foraging with commercial purposes entails contradictions that are hard to keep in the same frame. The idea of sustainability connects the berries to the growing, albeit partly aspirational, green economy of Arctic natural resources, while the commercial aspect keeps it closely tied with the capitalist logic and global market needs. In the context of the berry project, sustainability was used both to criticize and align with the capitalist logic. This resembles the use of the concept of progress in rooting mining in the region and the logic of parallelism underpinning it (see Chapter 3).

However, what made this assemblage unfunctional was the attempt to fit local jobs into the commercial berry assemblage. This insertion of state logic was not successful. As the berry project funding came from the municipality and EU periphery funding for economic development, the incentive was to create jobs for locals. If successful, this would tick two boxes: create new jobs in the region suffering from high unemployment and root the commercial berry picking in more sustainable picking practices by replacing migrant pickers by local pickers.

The local communal foraging practices set superberry foraging apart from the commercial mass berry picking done by migrant labour. Two factors became critical in terms of sustainable superberry foraging. Firstly, they must be carefully handpicked. Heli explains that the superberries preserve their nutritional super-qualities only when hand-picked to avoid bruising the berries and they need to be frozen immediately. Secondly, following the sustainability logic, the berries must be sustainably harvested both ecologically and socially. Heli envisions this happening as part of the berry project by involving locals in commercial berry picking. Local pickers would become the bodies needed for the commercial supply chain, but their localness and berry-picking practices would make the foraging socially sustainable. As a result, the berries could keep their superberry-value.

An August afternoon event where Heli gave this talk represented an attempt to revalue both the wild berries as superberries and their foraging as an attractive and profitable job for locals. Heli's talk about superberries was part of a public, widely advertised event about commercial Arctic berry picking, designed to attract more locals to participate in commercial picking with a promise of seasonal work and extra earnings. To root the berry project in both the local political economy of work as well as ideas of the green economy (Mendoza 2018; Norrgard 2008), the project targeted local people who might be interested in extra income: the retired or unemployed; young people; students and seasonal workers.

In the project's early days, Heli believed that a combination of awareness building, the use of communal practices, and an increase in price per kilo paid to the pickers would attract locals to engage with commercial berry picking. Heli imagined keeping the commercial picking rooted in communal foraging practices. She believed that consumers want not only a really healthy blueberry, but also green and sustainable development. Heli's translation of this niche market profile was to turn foraging into sustainable income for locals, making foraging a more sustainable business. She was brainstorming for a special Lapland fair trademark and practices that would guarantee ethical and sustainable foraging practices, including fair wages for foraging work. To avoid stepping over the boundaries of family berry patches, Heli had an idea to use local knowledge and new technology to create an app that would show the free berry patches. At a minimum, Heli imagined that there would be at least a five kilometres wide communal zone around the villages, which could be left alone by commercial picking. The key for Heli was to plan all of this together with local communities so that family berry patches would be left alone.

Heli's attempts can be seen as a way to bring together global trends and practices rooted in the green economy and sustainable development with the local political economy of work and labour in the Arctic periphery. Through discursive work, including fair trade mark and sustainability discourses, Heli was both trying to make sense of and justify the conversion of commons into commodities. Ultimately, these attempts can be seen as a de- and reterritorializing attempt to revalue berries and make them movable across cultural and geographical terrains—with a profit. In assemblage terms, Heli tried to fit three different logics—berries as commons, commodities, and jobs—into one assemblage without success.

The biggest hindrance to Heli's plans to find local pickers willing to engage with commercial foraging was what she called the "fuck-you-prices" paid to pickers. The payment per blueberry kilo varies from one euro to two euros. The compensation was considered too low compared to the hardness of the work and the working conditions; therefore, no local would do commercial foraging for that price. Heli calculated that the minimum price would be 3 euros per blueberry kilo. This would yield around 45 euros per day, which would be tax-free income and might be attractive enough to the locals. Heli anticipated that this would make superberries more expensive to the consumers than the berries picked by migrant pickers, but she believed that enough customers were willing to do this. She compared more expensive superberries picked by local bodies to the eggs by the free chicken—a little bit more expensive but an ethical choice for the consumers: "berries from free pickers" could work along the lines of a similar logic that, for example, fair trade or local handicrafts are branded.

This was the plotting moment with capitalism. Disregarding the enduring inequality related to migrant pickers and the kilo price paid for them, in the berry project's imaginary, the quality of the superberries, together with the growing pull from the markets for sustainable products, translated to the higher kilo price for superberry pickers in the form of the local bodies needed to guarantee the sustainable practices and local jobs. Therefore, getting the right kind of berry pickers became the central goal of the berry project. What was in the making was socially creating a local body needed for sustainable foraging under the green economy logic. The migrant bodies, mainly male with a background in rural physical work, were socially produced as bodies needed for harsh physical foraging under the extractive capitalist logic with a very low salary. This inequality in the making was not discussed. The pickers, super-values, and bodies are socially, symbolically, and physically produced in the berry patches and their rhizomatic connections to capitalism.

The locals were not impressed. In fact, they consistently refused to engage in commercial foraging. The commercial berry-picking event on that August afternoon was a failure. Heli had booked a big auditorium to facilitate a big crowd, but only six people turned up: a middle-aged woman who recently moved to the area and wanted to know more about the berry patches nearby; a currently unemployed waiter, who is looking for a possibility to earn some extra money, a local woman who is studying to become a nature guide, and a group of students with their teacher from a local sports college, looking for a chance to earn extra money.

The advertisements promised that we would go out to the forest to learn the trade of picking berries for commercial purposes, but heavy rain put a stop to this plan: berries cannot be picked for commercial purses when raining as they get wet and are easily mushed during transportation. Instead, Heli tries to educate us through a PowerPoint lecture about foraging superberries in a sustainable way. In the end, the teacher and the students were the only ones interested in future possibilities for commercial foraging within super berry practices. During the project's lifespan, Heli organized several similar kinds of events with similar disappointing results: the locals were simply not interested in taking part in commercial picking, super or ordinary.

One reason seems to be that the value locals gave to their berries did not match the market price. For locals, the berries are more than a source of extra cash. The berry project did not allow locals to retain control of the social and cultural relations of production that were intertwined with berry foraging. Reterritorialized superberry foraging within the terrain of capitalist structures and arrangements did not work since it did not match how people organized their lives and their berry picking. When there is no desire, there is also no convergence. The berry assemblages overlap but do not merge. As one of the locals who took part in the afternoon berry event described: "I first need to get my own two freezers full. And I also have my nephew and an aunt down in the south, and I have promised to pick some berries for them, too. I really only came here to hear if there is news related to foraging and just to see who came to this event."

After this failed attempt to reterritorialize foraging as paid work for locals, Heli developed a plan B to find suitable bodies for ethical and sustainable foraging. She started to plan a move from local pickers to lifestyle pickers by attracting "backpackers around the world." Heli speculated that berry foraging could be rebranded or recoded as a chance to see the world and the Arctic nature and make the most of the global trend of getting in touch with pure nature and local cultures. Heli believed that many would like to come to Lapland for two months to pick berries, and this is not impossible; it is just that nobody has yet done it. This was another moment working simultaneously with and against capitalist logic: Heli was trying to convert the low foraging salary into a unique nature experience and a chance to travel abroad, now by deterritorializing foraging from paid work and reterritorializing foraging as a nature experience, leisure time, and traveling. This idea is borrowed from various voluntary work exchange frameworks, such as workaway, and is widely used in various fruit, berry, and tree projects around the world, with reported problems related to sustainability and ethicality (Duru 2021; Sanchez 2013; Zlolniski 2019).

In the summer, after the failed local recruitment events, the project set out to test the use of holiday workers as berry pickers. However, this pilot also failed. A small group of people willing to do foraging as part of the working away experience were recruited to the project, and together with one project worker, they stayed at one closed village school and went foraging for several days. However, as the motivation for people doing work-away is more about traveling and cultural experience, the aspiration for foraging with quantities needed for commercial picking was not achieved. The work-awayers really wanted to have an experience in their own time and at their own pace (Peltola 2018). The question of finding local berry pickers for commercial foraging remains unsolved. After the short break related to the pandemic in the summer of 2020 and 2021, the migrant pickers from Thailand have returned to the forests of Lapland.

By engaging with superfood discourse and sustainable foraging practices, Heli tried to create an aspirational assemblage that would have led to new kinds of combinations and possibilities.

This was a high ambition goal, and it remained largely an aspirational attempt, given the several contradictions inherent to the idea as well as being embedded in a highly complex political economy of Arctic natural resources. While the employment of superfood discourse worked in the sense that the Arctic berries are revalued as superberries, revaluing foraging practices was more difficult. The berries are at the intersection of communing and commodifying, and while the locals were more open to the idea of discursively revaluing the mundane berries into superberries, they were resistant to changing their own foraging practices from commons into paid work. For them, the berries are not solely about economics but also about politics and culture, values that were not convertible at the time. In addition, the transformation from domestic subsistence to well-being is also a matter of class, and this particular transformation was not appealing to the locals. The discourse of the superberry is directed to urban consumers under hectic and stressful conditions, while the arctic conditions require another type of endurance, strength, and know-how. Berries are certainly part of it, but ordinary berries will do the job.

Moreover, commercial picking would have changed the relations of production and governance of berries, making the locals more dependent on global commodity prices and turning them into vulnerable pickers in a working economy where they are marginal and easy to replace. The local Arctic order of berries continues to firmly root the berries as commons under social governance, which seems to be a resistance to capitalization. At the same time, the migrant pickers are exactly in that kind of position that the locals reject: vulnerable, easily replaced and unskilled, working under poor conditions, underpaid for the global economy. The berry assemblage also shows how capitalist conversions are never total or complete, but rather capitalism always requires non-capitalist social relations to accomplish its goals (Tsing 2015). Revaluation offers a translation that makes it possible for berries to become malleable and moveable. But the transformation is ragged: rough edges—*lines of flight* in Deleuze and Guattari's terminology—open up a possibility for political and creative work.

4.5 Unfunctional Commercial Berry Assemblage

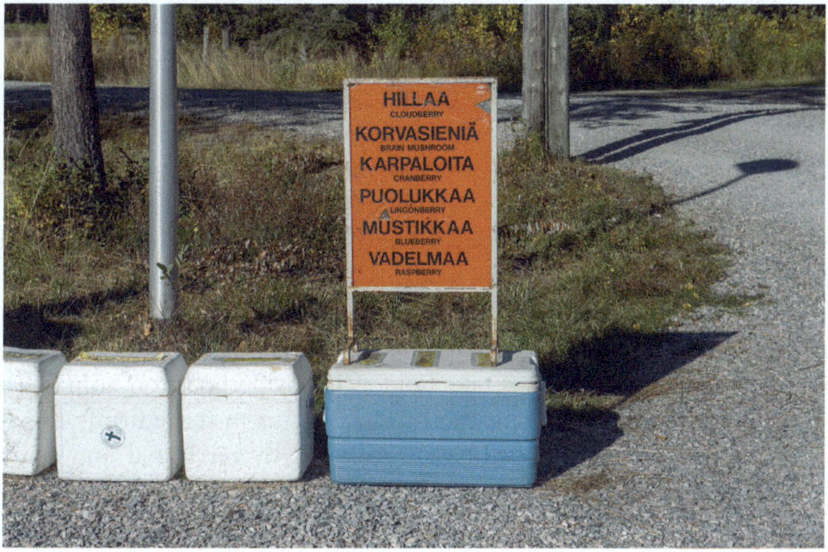

Fig. 4.5 Locally picked berries for sale

In 2024, critical actors in berry companies operating in the Finnish Arctic were convicted of human trafficking over the treatment of Thai berry pickers. The appalling working conditions of the imported berry pickers have been regularly discussed in the public domains, and there have been some attempts by various projects, artists, and activities to draw attention to the unsustainable practices related to commercial berry picking (see, e.g., Huhmarniemi 2024). The conviction recoded the commercial picked wild berries as blood berries, connecting them with blood diamond—a discourse which refers to human rights abuses, forced labour, illegal trade, and financing conflicts and wars related to the diamond industry (Yle Areena, accessed 1.5.2024).

The conviction accelerated the development of ethical and sustainable berry picking practices. For example, in 2004, Fair Trade Finland launched a project aiming to create local certification criteria for ethical berries to encourage the responsible production of domestic wild berries for the markets. The idea is quite similar to that of Heli's berry project, and most likely, it will also face the same problem: if the consumers are not willing to pay a considerably higher price for the berries, the wage for berry picking work will not be sufficient to attract local pickers (Fig. 4.5). The sustainable yet available and willing bodies for berry picking work are yet to be found.

Meanwhile, the Finnish media report a new concern: there may not be any Finnish wild berries to buy next summer in the shops as no berry pickers are available to do the work. Alternatives being discussed include buying exported wild berries from Poland or Latvia or foraging your own berries from the common berry fields.

The commercial wild berry assemblage still does not function properly, and perhaps it never will under capitalist order. Heli's attempt for sustainable berry assemblage did not take hold as it was perhaps too nomadic at the time. The wild berries and their utilization seem to be firmly rooted in the berry commons assemblage, organized by territorial and partly nomadic politics and aligning with more traditional gift exchange logics.

References

Boulanger-Lapointe, Noémie., José Gérin-Lajoie, Laura Siegwart Collier, Sarah Desrosiers, Carmen Spiech, Gregory H. R. Henry, Luise Hermanutz, Esther Lévesque, and Alain Cuerrier. 2019. 'Berry Plants and Berry Picking in Inuit Nunangat: Traditions in a Changing Socio-Ecological Landscape'. *Human Ecology* 47 (1): 81–93. https://doi.org/10.1007/s10745-018-0044-5.

Cavanaugh, Jillian R. 2007. 'Making Salami, Producing Bergamo: The Transformation of Value'. *Ethnos* 72 (2): 149–72. https://doi.org/10.1080/00141840701387853.

Dahlin, Johanna, and Martin Fredriksson. 2017. 'Extracting the Commons'. *Cultural Studies* 31 (2–3): 253–76. https://doi.org/10.1080/09502386.2017.1303428.

Deleuze, Gilles, and Felix Guattari. 1980/1987. *A Thousand Plateaus: Capitalism and Schizophenia*. Translated by Brian Massuni. London: Bloomsbury Academic.

Duru, Deniz Neriman. 2021. 'Workaway: Cultivating Conviviality within Mobility, Sociality and Daily Living'. In *The Routledge Handbook of Mobile Socialities*, ed. Annette Hill, Maren Hartmann, and Magnus Andersson, 238–50. London: Routledge.

Eriksson, Madeleine, Aina Tollefsen, and Anna Sofia Lundgren. 2019. 'From Blueberry Cakes to Labor Strikes: Negotiating "Legitimate Labor" and "Ethical Food" in Supply Chains'. *Geoforum* 105 (October): 43–53. https://doi.org/10.1016/j.geoforum.2019.07.003.

Fabinyi, Michael. 2013. 'Social Relations and Commodity Chains: The Live Reef Fish for Food Trade'. *Anthropological Forum* 23 (1): 36–57. https://doi.org/10.1080/00664677.2012.748645.

Feeny, David, Fikret Berkes, Bonnie J. McCay, and James M. Acheson. 1990. 'The Tragedy of the Commons: Twenty-Two Years Later'. *Human Ecology* 18 (1): 1–19.

Finnish Food Authority. 2020. 'Marsi 2019: Luonnonmarjojen Ja -Sienten Kauppaantulomäärät Vuonna 2019'. Report. Finnish Food Authority. https://www.ruokavirasto.fi/globalassets/viljelijat/tuet-ja-rahoitus/marsi-2019-raportti.pdf.

Foucault, Michel. 2014. *On the Government of the Living: Lectures at the College de France*. Edited by Arnold I Davidson. Translated by Graham Burchell. London: Palgrave.

Fred, Mats. 2018. 'Projectification: The Trojan Horse of Local Government'. Doctoral Thesis, Lund University.

Gershon, Ilana. 2018. 'Employing the CEO of Me Inc: US Corporate Hiring in a Neoliberal Age'. *American Ethnologist* 45 (2): 173–85. https://doi.org/10.1111/amet.12630.

Harvey, David. 1981. 'The Spatial Fix—Hegel, Von Thunen, and Marx'. *Antipode* 13 (3): 1–12. https://doi.org/10.1111/j.1467-8330.1981.tb00312.x.

Hedberg, Charlotta. 2013. '"Grapes of Wrath"? Power Spatialities and Aspects of Labour in the Wild Berry Global Commodity Chain'. *Competition & Change* 17 (1): 57–74. https://doi.org/10.1179/1024529412Z.00000000025.

Hedberg, Charlotta. 2021. 'Entwined Ruralities: Seasonality, Simultaneity and Precarity among Transnational Migrant Workers in the Wild Berry Industry'. *Journal of Rural Studies* 88 (December): 510–17. https://doi.org/10.1016/j.jrurstud.2021.04.008.

Hedberg, Charlotta, and Irma Olofsson. 2022. 'Negotiating the Wild West: Variegated Neoliberalisation of the Swedish Labour Migration Regime and the Wild Berry Migration Industry'. *Environment and Planning A* 54 (1): 33–49.

Hodgson, Damian, Mats Fred, Simon Bailey, and Patrick Hall, eds. 2019. *The Projectification of the Public Sector*. London: Routledge.

Huhmarniemi, Maria. 2012. 'Berry Wars: A Science Centre as a Forum for a Dialogical Activist, Interdisciplinary Art Project'. *International Journal of Education through Art* 8 (3): 287–303. https://doi.org/10.1386/eta.8.3.287_1.

Huhmarniemi, Maria. 2024. 'Berry-Picking Culture and Conflicts in Finland1'. *International Journal of Education through Art* 20 (1): 51–67. https://doi.org/10.1386/eta_00151_1.

Jensen, Anders, Christian Thuesen, and Joana Geraldi. 2016. 'The Projectification of Everything: Projects as a Human Condition'. *Project Management Journal* 47 (3): 21–34. https://doi.org/10.1177/875697281604700303.

Joks, Solveig. 2022. 'Frustrated Caretakers'. In *The Sámi World*, edited by Sanna Valkonen, Áile Aikio, Saara Alakorva, and Sigga-Marja Magga, 150–64. London: Routledge. https://doi.org/10.4324/9781003025511-11.

Kulonen, Ulla-Maija, Irja Seurujärvi-Kari, and Risto Heikki Pulkkinen. 2005. *The Saami: A Cultural Encyclopaedia*. Helsinki: Suomalaisen Kirjallisuuden Seura.

Lacuna-Richman, Celeste. 2021. 'The Seasonal Migration of Thai Berry Pickers in Finland: Non-Wood Forest Products for Poverty Alleviation or Source of Imminent Conflict?' In *Social-Ecological Diversity and Traditional Food systems. Opportunities from the Biocultural World*, edited by Ranjay Kumar Singh, Nancy J. Turner, Victoria Reyes-Garcia, Jules Pretty, 91–105. https://doi.org/10.1201/9781003246220-4.

Länsman, Anni-Siiri. 2004. '*Väärtisuhteet Lapin Matkailussa : Kulttuurianalyysi Suomalaisten ja Saamelaisten Kohtaamisesta*'. Doctoral Dissertation, Oulu University. Inari: Kustannus Puntsi.

Lehtola, Veli-Pekka. 2012. *Saamelaiset suomalaiset: kohtaamisia 1896–1953*. Suomalaisen Kirjallisuuden Seuran toimituksia 1351. Helsinki: Suomalaisen Kirjallisuuden Seura.

Li, Jitong, and Karen K. Leonas. 2019. 'Trends of Sustainable Development Among Luxury Industry'. In *Sustainable Luxury*: Managing Social and Environmental Performance in Iconic Brands, edited by Miguel Angel Gardetti and Ana Laura Torres, 107–26. Singapore: Springer Singapore. https://doi.org/10.1007/978-981-13-0623-5_6.

Loyer, Jessica. 2016. 'Superfoods'. In *Encyclopedia of Food and Agricultural Ethics*, edited by Paul B. Thompson and David M. Kaplan, 1–7. Dordrecht: Springer Netherlands. https://doi.org/10.1007/978-94-007-6167-4_574-1.

Mauss, Marcel. 1954. *The Gift. The Form and Reason for Exchange in Archaic Societies*. London: Cohen & West.

Mendoza, Marcos. 2018. *The Patagonian Sublime: The Green Economy and Post-Neoliberal Politics*. Rutgers: Rutgers University Press. https://doi.org/10.2307/j.ctvscxrwx.

Narotzky, Susana. 2013. 'What Kind of Commons Are the Urban Commons?' *Focaal—Journal of Global and Historical Anthropology* 66 (2013): 122–24. https://doi.org/10.3167/fcl.2013.660113.

Natural Products. n.d. Maa- Ja Metsätalousministeriö. Accessed 1 June 2024. https://mmm.fi/en/natural-products.

Nilsson, Lena Maria. 2022. 'Sámi Food Culture'. In *The Sámi World*, edited by Sanna Valkonen, Áile Aikio, Saara Alakorva, and Sigga-Marja Magga, 165–80. London: Routledge. https://doi.org/10.4324/9781003025511-12.

Norrgard, Chantal. 2008. 'From Berries to Orchards: Tracing the History of Berrying and Economic Transformation among Lake Superior Ojibwe'. *The American Indian Quarterly* 33 (January): 33–61. https://doi.org/10.1353/aiq.0.0035.

Olofsson, Irma. 2024. *Invisible yet Essential: The Role of Seasonal Labour Migration in Sweden's Green Industries*. PhD dissertation, Umeå University. diva2:1830027.

Olofsson, Irma, Aina Tollefsen, and Charlotta Hedberg. 2023. '"A STRONG MIND AND A SOLID PHYSIQUE": Symbolic Constructions of Migrant Workers in Sweden's Green Industries'. *Nordic Journal of Migration Research* 13 (2): 1–17.

Osteen, Mark. 2002. *The Question of the Gift: Essays Across Disciplines*. London: Routledge.

Peebles, Gustav. 2015. 'Anthropology and Capital'. *HAU: Journal of Ethnographic Theory* 5 (1): 477–79. https://doi.org/10.14318/hau5.1.021.

Peltola, Noora. 2018. *Selvitys Keruuverkoston Poimijoista*. Thesis: Lapland University of Applied Sciences.

Radovic-Fanta, Jelena. 2021. 'Good Mothers and Good Workers: Discipline and Care in Chile's Grape-Packing Plants'. *Anthropology of Work Review* 42 (1): 3–13. https://doi.org/10.1111/awr.12216.

Rantanen, Pekka, and Jarno Valkonen. 2011. 'Ulkomaalaiset metsämarjapoimijat Suomessa'. Report. Ministry for Foreign Affairs Finland.

Reisman, Anne. 2019. 'Luonnontuoteala'. Report. Ministry of Economic Affairs and Employment of Finland.

Reyes, Angela. 2014. 'Linguistic Anthropology in 2013: Super-New-Big'. *American Anthropologist* 116 (2): 366–78. https://doi.org/10.1111/aman.12109.

Sanchez, Teresa F. 2013. 'Californian Strawberries: Mexican Immigrant Women Sharecroppers, Labor, and Discipline'. *Anthropology of Work Review XXXIV* 1: 15–26.

Shankar, Shalini, and Jillian R. Cavanaugh. 2012. 'Language and Materiality in Global Capitalism'. *Annual Review of Anthropology* 41 (1): 355–69. https://doi.org/10.1146/annurev-anthro-092611-145811.

Siitonen, Pauliina, Mirka Rauniomaa, and Tiina Keisanen. 2021. 'Language and the Moving Body: Directive Actions With the Finnish Kato "Look" in Nature-Related Activities'. *Frontiers in Psychology* 12 (June): 661784. https://doi.org/10.3389/fpsyg.2021.661784.

Sikka, Tina. 2019. 'The Contradictions of a Superfood Consumerism in a Postfeminist, Neoliberal World'. *Food, Culture & Society, 22* (3): 354–375. https://doi.org/10.1080/15528014.2019.1580534.

Spangen, M., A.-K. Salmi, T. Aikas, and V.-P. Lehtola. 2015. 'Sami Histories, Colonialism, and Finland'. *Arctic Anthropology* 52 (2): 22–36. https://doi.org/10.3368/aa.52.2.22.

Teetl'it Gwich'in Renewable Resources Council, Brenda Parlee, Fikret Berkes, and Teetl'it Gwich'in Renewable Resources Council. 2006. 'Indigenous Knowledge of Ecological Variability and Commons Management: A Case Study on Berry Harvesting from Northern Canada'. *Human Ecology* 34 (4): 515–28. https://doi.org/10.1007/s10745-006-9038-9.

Trusler, Scott, and Leslie Main Johnson. 2008. '"Berry Patch" As a Kind of Place—the Ethnoecology of Black Huckleberry in Northwestern Canada'. *Human Ecology* 36 (4): 553–68. https://doi.org/10.1007/s10745-008-9176-3.

Tsing, Anna Lowenhaupt. 2015. *The Mushroom at the End of the World: On the Possibility of Life in Capitalist Ruins*. Princeton: Princeton University Press.

Turtiainen, Marjut, and Pekka Rantanen. 2020. 'Onko mustikoiden ja puolukoiden poiminta Suomessa tehostunut thaimaalaisten poimijoiden myötä?' *Maaseutututkimus* 28 (2): 61–82.

Tuunanen, Pekka, Markus Tarasti and Anne Rautiainen. 2012. Eds. *Jokamiehenoikeudet Ja Toimiminen Toisen Alueella. Lainsäädäntöä ja hyviä käytänteitä.* Suomen Ympäristö 30/2012. Ministry of the Environment of Finland. Helsinki: Edita.

Veloz, Paula. 2020. 'Value Chains and Soft Commodities in Amazonia. Regulatory Prospects for Commodified Biodiversity According to the Glocal Production Chain of Açaí'. *Journal of Land Use Science* 15 (2–3): 389–405. https://doi.org/10.1080/1747423X.2019.1694663.

Yan, Yunxiang. 2012. 'The Gift and Gift Economy'. In *A Handbook of Economic Anthropology, Second Edition*, edited by James G. Carrier. Cheltenham: Edward Elgar Publishing. https://doi.org/10.4337/9781849809290.00025.

Yle Areena. *Yle Lappi Uutiset*. Accessed 1 May. 2024.

Ylikoski, Sofia. 2024. *Ihmiskaupan Ilmentyminen Suomessa*. Thesis. Vaasa: Vaasan ammattikorkeakoulu. https://urn.fi/URN:NBN:fi:amk-202404106200.

Zlolniski, Christian. 2019. *Made in Baja: The Lives of Farmworkers and Growers behind Mexico's Transnational Agricultural Boom*. Berkley: University of California Press. https://doi.org/10.2307/j.ctvp2n2hw.

5

Northern Lights Assemblage: Making a Living Out of Thin Air

5.1 Dancing Lights on the Sky

Fig. 5.1 Aurora Borealis

S. Pietikäinen, *Cold Rush*, Language and Globalization,
https://doi.org/10.1007/978-3-031-63995-1_5

> The thick overall feels tight around my body, and the heavy winter boots slow me down. My movements are clumsy, and I can feel the woolen socks pinching my toes. I start sweating. I look around at my co-travelers, who look like astronauts, all wearing similar uniforms: black overalls with reflective strips, big winter hats and boots, serious gloves, a head torch, and a professional camera attached to a pod in one hand. We have just finished a crash course in Aurora photography with our guide, Matti, a professional photographer and tourism entrepreneur specializing in Aurora chasing. Beautiful pictures of Aurora on the wall of Matti's studio heighten the anticipation of the real possibility of seeing the lights tonight. With a high price tag attached to the hunt, expectations run high. We are set to go for a Northern Lights hunt, a guided 4-hour tour in Matti's van through the dark Arctic night fueled with the hope to spot one of the Arctic nature wonders and catch it with our cameras. We follow our guide through a dark, cold December night near a skiing resort in the Finnish Arctic. It is minus -25 Celsius and pitch dark. We are settling into a van, and our hunt starts. (Fieldnotes, January 2017)

Aurora Borealis, also known as Auroras or Northern Lights in the Northern Hemisphere, and Aurora Australia and Southern Lights in the Southern Hemisphere, is a celestial nature phenomenon visible to the human eye on dark polar nights (Fig. 5.1). These dancing lights have intrigued the human imagination across centuries for their believed magical or mystical properties as well as their spectacular beauty. Auroras, like reindeer, Santa Claus, and huskies, belong to the bucket list for many international tourists visiting the Finnish Arctic, making them a key element for developing winter tourism (Bertella 2013; Jóhannesson and Lund 2017; Weaver 2011). Also, in Finland, Auroras have become a thriving business (Heimtun et al. 2015), and by 2023, they have become the most popular activity among international tourists (Visit Lapland, accessed 15 January 2023).

I noticed the first signs of Aurora tourism while conducting fieldwork in the Finnish Arctic in the early 2000s. I took a taxi from the airport to my destination; a remote village, and the driver told me about international tourists who wanted to hire a taxi and "just drive around" in the dark night, hoping to see the Northern Lights. Later in the week, in the evenings, I noticed a swarm of tourists gathering in front of the hotel, armed with hot flasks, cameras, warm overalls, and folding chairs. They peered hopefully up at the sky and went inside the lobby to warm up at regular intervals while the rest of us watched them from the comfort of our rooms or dinner tables.

Since these early attempts often organized by the tourists themselves, Finnish Aurora tourism has grown into a professional business. The basic idea behind the Aurora hunt is simple: the tourists are taken on a car drive through the dark Arctic night looking for the Northern Lights and to take pictures of them. At a minimum, only a very basic infrastructure is needed: "A man in a van" as one local Aurora tourist entrepreneur described the core of the business. In reality, however, the Aurora hunt is a complex tourism product to produce and manage successfully: Auroras are unpredictable, ephemeral nature phenomenon that require specific weather conditions for the human eye to see. Weather conditions in the Arctic nights can be cold and snowy, creating a need for high-quality gear and good driving skills. However, the biggest challenge is managing expectations: what to do while waiting for the Auroras to show up, and how to handle the disappointment if they are not seen?

The sun's activity creates Auroras. When solar winds reach Earth, charged particles collide with atoms in the high-altitude atmosphere, producing lights in an oval-shaped region around the planet's magnetic poles. The best geographical location for seeing Auroras is near the poles. In the North, this is called the *Aurora zone*, a belt between the 70° and 65° contours. Most of the Finnish Arctic area is located in this zone, including the villages and tourism resorts that I have been researching Cold Rush.

Auroras take different shapes and colours, but the most common Aurora displays are in a folding curtain-like form in greenish lights that dance slowly across the sky. Rarely, they can appear in deep purple and red in a circular shape. To see Auroras with your own eyes requires a clear sky and a dark night. In the Finnish Arctic, this means that the Auroras can only be seen between September and April, after which the white nights of the Arctic summer erase the Auroras for the human eye. Also, light pollution diminishes the visibility of Auroras, making the peripheral areas of the Finnish Arctic perfect for Aurora hunting.

The lure of Auroras chimes with beauty and mysticism. The scientific and natural wonders of the Auroras are enhanced by traditional stories and beliefs, providing explanations for their origin and meaning. The Latin name, used in European scholarly literature since the seventeenth century, *Aurora Borealis* combines both the mythical and geographical aspects of the Auroras: Aurora refers to the Roman goddess of dawn, and Borealis means Northerly (Falck-Ytter et al. 1999). The Finnish Aurora zone hosts multiple Aurora folklores. According to indigenous Sámi mythology, the Auroras were spiritual beings who could hear and understand people (Helander-Renvall 2006: 27). The Finnish folktale describes them as Fox fires (Auroras are called in Finnish *revontulet*, meaning fox's fire): When the fox's tail touches the mountains in the North, sparks fly into the sky, igniting the Lights. The Aurora sightings have been interpreted as both a good and bad omen (Brekke and Egeland 2012). I remember my mother telling a story about her childhood in eastern Finland during the Second World War. The Auroras were so strong one winter that they painted a red glow on the white snow. People interpreted this as a sign of a great loss of lives. While doing fieldwork in the Arctic region, I met tourists who travelled to the Finnish Arctic as they believed that seeing the Northern Lights had the power to cure their illnesses, bring them luck in their marriage, or increase fertility. Regardless of the stories or reasons to see Auroras, their beauty and the fact that they come from far away, from space, triggers our senses and imagination.

Against these vivid stories about the Auroras together with the technologically enhanced, photoshopped pictures of Auroras commonly used in tourism, the real-life experience may pale in comparison. This is what happened to my co-researcher, Alexandre Duchêne. We worked together on the Cold Rush project, and despite his several visits to the Finnish Arctic, he had never seen Auroras. However, one December night, when we were working in a well-illuminated ski resort, the lights danced above it, and Alexandre saw them for the first time. But this was a bit of anticlimax as they were not what he had expected. This is what he wrote in his field notes:

> I was very surprised. I saw some greyish moving traces in the sky but I never would have thought they were The Northern Lights. I have been visiting the region regularly for more than four years, and I was always joking with the fact that I had never seen the Northern Lights and I was teasing Sari that her Northern Light might well be a myth she sold me. For sure, I was surrounded by images of the Northern Lights in flyers, posters, and advertisements, depicting huge greenish or reddish immense flux of lights. That was the Northern Lights for me, my myth! What Sari showed me was very different from what my constructed images of the Northern Lights were. (…) Later, while going to the restaurant, Sari again pointed to the sky and indicated the Northern Lights. It was stronger but still greyish, not greenish. However, I was able to see it better, as if my eyes and my brain were accepting that the Northern Lights looked different from my imagination or how the images I saw had socialized me into believing what they looked like. (Alexandre's fieldnotes, December 2019)

Alexandre's experience illustrates how seeing the natural Northern Lights can actually be disappointing. Commercial and popular culture-enhanced representations of the Northern Lights construct a pre-conceptualization of how they should look, and the reality may be disappointing. His experience also shows how experiencing Auroras benefits from management: someone who can point to it, tell a story about it, change your ways of seeing it, and has the authority to confirm that the vibrant lights are indeed the Northern Lights. This need for expertise creates a niche market for guided Aurora hunting products.

In this chapter, I examine how the so far un-commodified Arctic commons, the Northern Lights, have become a popular tourism activity and a booming enterprise in the Finnish Arctic with the help of critical assemblage analysis (see Chapter 2). This revaluing is situated in a rhizome of four assemblage politics: the combination of *territorial* and *capitalist* assemblages allows the revaluation of the Auroras as a post-tourism product for global markets, only available in the Arctic. However, locally Aurora product functions within *state* tourism assemblage, which creates tensions for global business practices and calls for *nomadic* strategies. Moreover, the ephemeral nature of the Northern Lights means that the Aurora product needs constant reassembling to facilitate the changing weather conditions and the needs of tourists.

Assembling becomes a key task and critical skill in Aurora business. As we shall see, this translates into constant de/reterritorializing and de/recoding work.

I focus revaluation of the Northern lights through a longitudinal ethnography with Aurora business pioneers in the Finnish Arctic, Lucy and Matti, who were among the first entrepreneurs to build their business around Auroras. For over six years, I followed how their Aurora business developed from scratch into a successful international tourism business, even surviving the tourism bust caused by the global COVID-19 pandemic, while they themselves became gradually rooted as local residents with expanding connections to the local community.

Lucy and Matti moved to the Finnish Arctic with their family in search of a lifestyle change after living an urban life outside of Europe. They were looking for a terrain where they could transform their academic training in Arts, Social Sciences, and photography as well as their cultural capital as cosmopolitan English speakers into a viable income. They chose Finland primarily due to Matti's heritage as a second-generation Finn with Finnish skills and family connections. However, the decision to pursue the Aurora hunting business resulted from multiple factors. They were looking for a business idea that was not completely saturated with competition, which would not require substantial financial investments. Being English speakers and users of Finnish meant that they targeted international tourists visiting the Arctic. The idea of Northern Lights hunting came out of brainstorming with local tourism development agents, and they received start-up funding as part of a regional economic development initiatives. (see Chapter 3). Lucy and Matti were the first entrepreneurs in the region to make Aurora hunting their primary product and to earn a successful living from it.

Their story shows the tensions and innovations involved in the economic transformation of both Arctic natural resources and academic skills into a successful Aurora business. As newcomers, Lucy and Matti see Auroras differently. Their out-of-the-local-box approach, *nomadic* in Deleuze and Guattari's (1980/1987) terminology, connects their Aurora business to the global, cosmopolitan terrain of post-modern Arctic tourism beyond the local, nation-state tourism infrastructure and traditional practices. Their business transgresses several traditional tourism boundaries and practices. Most of the time, these changes unfold smoothly, but at times, they get stuck in frozen national and local patches with fixed norms and practices, creating tensions that Lucy and Matti have experienced and have had to find strategies to navigate around.

Their story shows the intensity of assemblage work requires to successfully transform the common into a commodity. Their work entails constant assembling work—selecting, fitting, and organizing a functional Aurora hunt—which is needed every night to accommodate the desires of a new group of tourists within the changing weather conditions. Their story also unfolds the enduring navigation across the state logics framing the local resources and belongingness, and the intensity of the reterritorialization and recoding of their products, skills, and trajectories to fit simultaneously both to the local, nation-state-organized tourism assemblage as well as to global, high-end post-tourism flow. Their story also shows how the global pandemic impacted the entrepreneurs' lives, particularly where the sector depends on international tourists, and the survival strategies they developed.

The breadth and depth of the research collaboration with Lucy and Matti have allowed me to follow how their business evolved and developed. For six years, I took part in their programs, observed them at their work, and met regularly with them to discuss their business. We also talked about the different strategies they developed to manage the seasonality of Arctic tourism, the constraints between national and postmodern tourism frames, and the local political economy of authenticity and belongingness. We have also shared reflections about the Cold Rush. I also followed their media presence, including online reviews by their guests. Lucy and I stayed in touch during the pandemic, which is reflected in this chapter's focus on her experience. During this time, we also became neighbours on the Arctic scale as we live within 100 kilometres of each other when I am staying at our second home in Lapland.

5.2 Aurora Buzz: From Frozen National Patches to Global Post-tourism Flow

Fig. 5.2 Hunting for the Northern Lights

I feel the car slowing down and finally stopping. Is this it? Are the Auroras dancing in the sky? We had been tracing them across the dark Arctic winter night for over an hour with no sightings, and I was already feeling a bit sleepy from the steady movement of the car. Our guide, Matti, parks the van near a frozen lake with a view of vast open skies. We climb out of the warm car on a chilly night, and I am glad I am wearing the warm overalls, protecting me from the cold. I peer out into the darkness, and at first, I see nothing. Gradually, my eyes adjust to the darkness, and I spot familiar star constellations: the Big Dipper, the Milky Way, and possibly the Northern Star. Matti examines the Aurora forecast app on his mobile and declares there is a high probability of seeing the Auroras within the next few hours, but we may need to move to another place for a clearer sky. I feel a growing excitement: The possibility is there! Even though I have seen Auroras before, they are always spectacular. We briskly unload

> our hunting gears from the car, following Matti's direction, and set our cameras on pods facing the Northern sky. We are ready. The waiting prowling begins. (Fieldnotes, January 2018)

When Lucy and Matti started their Aurora Business in the mid-2010s, they were pioneers (Fig. 5.2). The Aurora business was already booming in many other Arctic tourism destinations, including Alaska, Iceland, and Norway (Heimtun and Haug 2022), but had not yet reached Finland. Recoding the free natural phenomenon into a tourist product with a price tag was initially resisted locally. For many locals, the idea of paying to see the Northern Lights seemed far out and unimaginable at best and dubious at worst, similar to Heli's attempts to turn berry picking in sustainable local work (see Chapter 4). Lucy's introduction of their business idea at a local tourism development meeting was met with suspicion. She recalls how one participant asked them directly how they could promise something that they could not deliver for sure. At that time, the idea of non-commodifiable Auroras was deeply rooted in local tourism practices. When I inquired at a local hotel reception about the possibility of an Aurora hunting product, she was puzzled and told me the following instead: "You can just step outside and peer into the skies. If Auroras are there you will see them, there is no need to pay for that. It is free".

The suspicion stems from a tension between nation-state-organized traditional Lapland tourism and post-modern Arctic tourism, which Lucy and Matti represent. From the assemblage perspective, the tension stems from the difference between two distinct flows and codes. The critical distinction is indexed already in the naming of the territory: Lapland is a geopolitical name for the Northernmost county of Finland, while the Arctic is a more of an ambivalent term, having a geographical reference point but also referring to a particular space. Moreover, Lapland is a code that indexes the nation-state assemblage, while the Arctic belongs to the global tourism assemblage.

Lapland tourism emerged in the early twentieth century as a way to perform citizenship for the newly independent Finnish nation. Lapland became constructed as a national site of the periphery that belongs on citizens' bucket list and needs to be visited for hiking or skiing as part of the Finnish citizenhood imaginary (Ren and Abildgaard 2021). In this state assemblage flow, Lapland nature was coded as an untouched nature that every Finn could enjoy under the historical common law, known as every person's right (Hautajärvi 2014). The key was that everything was free: to ski, hike, forage, fish, put a tent up, and explore Lapland's nature. This code was further reinforced by the potential to develop its peripheries economically. In this system, tourism is intertwined with local political and economic infrastructure, often financially supported by various EU, state, and municipality-funded projects (like the berry project, see Chapter 4). Investments in infrastructure, such as airports, railways, and roads, align with the nation-state's logic, keeping peripheries inhabited while expanding tax revenues. This often resulted in alliances between local tourism businesses and municipalities and the state.

As newcomers to the region, Lucy and Matti see Auroras and the Arctic differently. They think out of the local Lapland box. They connect Arctic tourism and Aurora business as part of a global flow of nature-based experience tourism, where nature's wonders and destinations are coded in terms of their uniqueness. This kind of post-modern tourism is organized in an attempt to create unique experiences in extraordinary places that still maintain ties with locals and are not yet overtly impacted by mass tourism (Jansson 2018). As customers, post-modern tourists have an appetite for exclusiveness, speculation, the extraordinary and the fabulous while also keeping an eye on sustainable sensibilities. The service providers have a crucial role as facilitators and cultural interpreters: they have insider access to the local histories and practices and can thus provide tourists with unique access to invisible connections and participation in local practices. A journey of discovery in the comfort of excellent service. This is the terrain on which Lucy and Matti root their business.

Aurora hunt plays well into the post-modern tourism flow: it is an untouched nature wonder that can be sensorially experienced in an unpopulated dark place, with connections to local histories and practices anchored by stories. Its ephemerality enhances it's allure by intensifying the experience (Harvey 1989). At the same time, post-modern Arctic tourism sits uncomfortably with the Lapland nation-state-driven tourism infrastructures and ideologies. It bypasses the local gate-keeping practices and established norms by reaching out directly to the niche, global market and international tourists. Lucy and Matti's cosmopolitan trajectory, university education, and working experience in cultural industries, together with their English skills and expertise in visual and oral narratives, locate them within this global flow and enable them to unfold it successfully. Globally, it makes them successful, but locally, it creates tensions, with the potential to position them as outsiders and question their legitimacy.

The tension proves to be an enduring one: Lucy struggles with it every step of the way. "I sometimes feel like the locals think I am this crazy foreign woman with crazy ideas." Their post-modern Aurora business positions them outside the local stratified terrain, embedded with norms, historical belonging, and traditional ways of doing things. She sometimes wonders if she has come "too early" as the local tourism industry appears "not ready" for her ideas. In assemblage terms, Lucy is locally presenting nomadic thinking outside local structures and norms (Deleuze and Guattari (1980/1987: 53–54). The ambivalence stems from multiple positions; she wants to belong to the local community but do business in a novel way. Lucy tries to solve the tensions in multiple ways: breaking free, stepping back, and finding an alternative way in: she volunteers at the local library, brings cookies to meetings, and participates in local sports. She tries to find a way in.

Apart from struggling for local legitimacy, Lucy and Matti have also struggled to find the right version of the Aurora chase, one that would work both for them and their customers. At the beginning of their business development, they imagined the product would revolve around learning to photograph Auroras with a professional photographer. But they quickly realized that only very few tourists were that keen on the professional aspects of night photographing per se. Rather, most tourists wanted to see the Northern Lights with their own eyes and take a photo as a visual souvenir of the experience.

This realization prompted Lucy and Matti to redesign their Aurora product. As in many other places in the Arctic, they decided to organize it around the concept of a chase. This move was a game shifter. It deterritorilized Auroras from the terrain of nature photographing and reterritorilized them into the terrain of a carefully designed nature tourism experience. This terrain was already firmly rooted in the tourism imagination of Arctic destinations, with untouched, unique nature as one of the biggest pull factors. It also freed Lucy and Matti to innovate and imagine a product that would go beyond Matti's photography skills and economically support them both.

The key recoding of this move was naming the product as an Aurora *chase*, which lends a sense of a hunt, a pursuit of an object to capture and consume. Importantly, this recoding frames the product in terms of action: rather than just seeing the Auroras, this product includes a series of activities, such as tracing, finding, and capturing Auroras. In practice, the Aurora chase includes gearing up, driving around, taking pictures at night, snacking, and chatting.

Coding the Aurora product as a chase expands the scope of the product's potential producers and consumers. Aurora hunting is no longer connected only to Matti's skills as a professional photographer, nor is it targeted at a specific niche market of photography enthusiasts. Rather, it becomes possible for both Lucy and Matti to deliver the hunting product in such a way that attracts a wide range of tourists: Aurora fans, families and groups looking for activities, and executive clients seeking team-building activities. The revenue prospects increased considerably.

The chase code also helped navigate two main local critiques related to fraud and inauthenticity. Putting a price tag on viewing Auroras was locally criticized as being a scam. Lucy adamantly rebutted the accusation by arguing that their business was not a rip-off as they did not promise people something that they could not deliver. In Lucy's experience, many locals refused to believe that space weather, including Aurora activity, could be predicted like any other weather condition. Lucy and Matti have become experts in space weather forecasting and, therefore, are confident to advertise with the slogan "We have a ninety-six percent success rate." Moreover, there is a niche market of international tourists willing to pay for the added value of turning the simple act of looking into the skies into a tourism activity of hunting Auroras, with excellent service and equipment (Heimtun and Lovelock 2017).

Rather than relying on traditional or local knowledge, often inaccessible due to lack of sufficient Finnish skills or access to local family traditions, Lucy and Matti rely on their academic training to find the resources necessary to figure out where and when to see the Northern Lights. They have become professional "Aurora finders." They also turn their non-local, cosmopolitan trajectory, and academic training into a resource to "escape the trap of authenticity." As they do not have generations-long family ties to the region, they cannot claim authenticity in terms of tradition, belongingness, and kinship as many other local entrepreneurs do in their marketing and product development. Matti and Lucy cannot tell stories about their grandparents' forests, handicrafts, or cooking to create a connection between the place, themselves, and their product. Instead, they move away from the authenticity code to a *genuine* code to describe their product. In the local political economy, authenticity is connected to the generations-long roots in the area, intertwined at times with indigenous Sámi connections (Pietikäinen et al. 2016). Instead, genuine is an open code that is available for anyone, including non-locals, as it constructs the product through its quality and intentions. It is close enough to authenticity for Lucy's purposes but without the political bloodline associations or links to her previous professional life. For Lucy, genuine is easier: "I visibly cringe when I hear the word authenticity."

As a tourism activity, Lucy compares Aurora chasing to fishing: the goal is not to catch a fish but rather to enjoy the wholesome, temporal process of experiencing fishing in nature. Fishing, like Aurora chasing, entails preparation, fishing, and enjoying the experience afterward. The success of a fishing trip is determined not only by the number of fish caught but also by how great the experience was. This effect is what Lucy and Matti set out to design: to assemble elements that would guarantee a satisfying experience of hunting Auroras, even when they are not seen, both experientially for the tourists and financially for Lucy and Matti.

5.3 "Nothing Bling but Something Delicious with Comfort": Designing Aurora Assemblage

Fig. 5.3 Fitting Aurora assemblage elements

> No show. I am blinking at the dark sky, no signs of Auroras. Our guide walks up to me, and gently guides and encourages me to take pictures and helps with technology (my battery died). Taking night pictures is more fun than I expected. After a while, the rush fades away, and the coldness settles in and bites my toes and fingers. I begin to feel restless and tired. One by one, we move to the warmth of an open fire. We are served hot coffee and tea along with the signature cloudberry brownies. We exchange stories and experiences with the Northern Lights. This becomes an experience of itself: sharing hot drinks and brownies by an open fire in the middle of the Arctic Night while listening to each others' stories about food, travels, and life. (Fieldnotes, January 2018)

Lucy and Matti set out to design a genuine Aurora chase with superb service and professionalism (Fig. 5.3).

They converge their skills, resources, and market niche into a wholesome, all-senses experience of Aurora chase, packaged with the comfort of excellent service: "Nothing bling but something delicious with comfort", Lucy summarizes. However, this convergence requires constant calibration of the best combination of key elements, partly due to changing weather conditions, partly to adapt to the particular needs and desires of the tourist group of a specific hunt. Good service and comfort are relational concepts. The perpetual designing of the Aurora chase product can be compared to crafting a vibrant kaleidoscope with permanent elements. However, the elements are organized slightly differently for each evening depending on the particularities of the hunt, weather conditions, the composition of the tourist group and their wishes. This assemblage work—fitting and arranging various elements—requires ongoing experiments and enduring navigations across Lucy and Matti's resources, the local political economy of tourism and natural resources, and international tourism markets. Assembling becomes their key job: as the conditions change, so too does the hunt change.

This continuous assembling and recalibration result from two intertwined forces: the rhythm of capitalism and the intensity of post-modern tourism. The capitalist mode of production for niche post-modern tourism markets requires flexibility not only in terms of content and working conditions but also in terms of pace. In the context of Aurora tourism, this creates a unique sense of intensity, a combination of acceleration (the hunt), and a frozen moment (when Auroras are seen and captured with cameras) (cf. Snyder 2016; Vandenberghe 2008). While constant calibration is central to Lucy's and Matti's work, the Aurora chase also includes more permanent material, affective, and discursive elements.

The core material elements of the Aurora chase assemblage include the Arctic landscape, darkness, access to the open sky, a van, snacks, the guide, and bodies equipped with warm clothes, cameras, and headlamps. Some of these elements are absolutely essential for the hunt to even be planned. The hunts cannot be organized during the white Arctic summer nights as the Auroras can only be seen in the dark. Without a car, it is impossible to drive outside of the light pollution of villages and tourist resorts or to chase Auroras across the terrain in search of clear skies.

At the same time, some material elements are deliberately chosen or left out for strategic distinction, in line with Lucy's and Matti's resources and vision of their product. They provide customers with professional-level photography equipment and winter-proof clothing. This codes the Aurora chase with two flows, professional photography and safety, both up-scaling the product. Professional camera equipment and a souvenir card as a farewell gift are tools for distinction on several levels. They distinguish this hunt from casual photographing with one's smartphone, rooting the hunt in the professional skills of this firm and creating conditions for better-quality photos as visual souvenirs of the hunt. Investment in professional-level Aurora chasing gear adds value to the Aurora chase as a product delivered by professionals, putting it at the higher end of tourist products on the local scale. Providing warm overalls, boots, gloves, and a headlamp lowers the participation threshold, allowing tourists to enjoy a night out in the Arctic without having to bring their own gear. This equipment also justifies the product's higher price compared to a free walk around the grounds of your hotel and peering into the sky with your mobile phone camera ready in your hand.

The final material touch to recode the hunt as a high-end, genuine Arctic product is offering guests homemade Arctic berry juice and cloudberry cookies made by Lucy from the berries she picked around the Aurora chase terrains. Instead of traditional sausage grilling, Arctic berry snacks create several important connections between the hunt region and Lucy's and Matti's business. The berries root Lucy into the terrain and position her as an expert with special access to and understanding of the land. The berry snacks provide food for her narratives about local traditions, the Arctic climate, and nature (see below). The berries also index sustainability, are nature-based, and offer a new experience for many international tourists to eat wild berry products. Resonating with post-tourism rationale (see e.g. Xiang et al. 2021; Jansson 2018; Ren and Abildgaard 2021), the berries enable the coding of the hunt as real and rare, with the affective impact of a genuine and unique experience. The berry cookies become a huge success, highly praised in the reviews, and ultimately, a signature treat of the hunt.

To enhance the affective aspect of the hunt, Lucy and Matti assemblage a product that tickles all the senses. The hunt is an embodied multisensory experience by default, as we sense the environment with our whole body (Ingold 2021). During the hunt, you feel the cold on your skin, creeping in through the gloves and winter boots, the sensation of breathing in the cold air and exhale evaporating, the warmth of the car, and the fire. Frozen nights also have a unique audioscape: the sound of snow under your feet, the sound of flowing water in the half-frozen river, or the snow-muffled voices of the others chatting while waiting for the Auroras. The frost has a distinct smell that blends with the fumes of smoke from the open fire and the scents from the surrounding forest. The flavour of the hot berry juice and cloudberry cookies by the fire add to the sensation of tasting the Arctic.

However, sight is the primary sense for experiencing the Auroras, making photography and Aurora chasing a perfect match. Lucy explains, that Auroras can be seen more clearly and in a broader spectrum through a camera lens than with the bare eye. Investing in professional cameras pays off in terms of increasing the chances of seeing Auroras, even if in a technologically mediated way. Even when the Auroras do not light up, the Arctic night can be a new sensation for those who come from over-illuminated, light-polluted landscapes: the dynamics of darkness and light offer an essential sensorial experience (Edensor 2013; Sun and Lv 2021). Against the dark Arctic night, you see stars and the moon, a faint glimmer of snow covering the trees and the land, flickering lights of passing cars or houses here and there. To tame the darkness, tourists are also given headlights that create rays of light in the night. Or they can enjoy the soft flickering light of an open fire.

The interaction between material and affective aspects of the hunt produces an ephemeral *Aurorascape*. Echoing the concept of affective landscape (Edensor 2013; Ingold 1993; Milani et al. 2019; Rinne 2021; Stroud et al. 2019), Aurorascape refers to a liminal, partly imaginary, partly geographically delineated space in which the Aurora chasing activities make sense. It points to a speculative Aurora chasing ground intertwined with a playful tourism playground. It is a terrain needed for de- and reterritorialization dynamics: transforming Auroras into embodied experiences, moving tourism bodies across the territory and involving them in various activities, and creating connections between locations, activities, and bodies. In the Aurora hunt, Aurorascape helps in producing an atmosphere, which is distributed through sensing bodies while also being palpable, a quality of environmental immersion (McCormack 2008).

The affective Aurorascape is designed to enable and intensify the embodied sensorial experience. Aurorascape aligns with the vitalist idea of emphasizing affects as processual and emergent (Katila et al. 2023; Wylie 2007). Following the vitalist stance in the work of Deleuze and Guattari (1980/1987), affect refers to a force that triggers reactions, becomes a force for action, or intensifies that action. Rather than simply being connected to subjective feelings or personal inclinations, affect can be understood as a sensation that involves not only subjectivity but also an affective field that connects the subject's body with other living and non-living force fields (Mortensen 2017: 5). In the hunt, affect highlights the transpersonal relations, both human and nonhuman, between the different elements within the Aurora assemblage, providing a broader, more-than-human concept of the social. Affect is generated by immersion in an Arctic environment that folds subject and space together (Anderson 2009: 78). Speculation around whether or not the Auroras will appear adds to the affective atmosphere of the hunt. Affect is crucial for creating the right mood for the hunt, the energy to engage, to take part. Affects, as intensive forces, are critical for our capacity to act. Matti and Lucy need to trigger such intensive forces: creating the mood for hunting, playfulness, adventure, prepping for readiness. The mood is vital as it impacts how we engage with the tourism prompts (Gonçalves 2020; Tucker and Shelton 2018); the mood has to be created, intensified, and discursively constructed and maintained by the hosts.

Ultimately, the Aurora chase is a discourse job. Turning the expectations, uncontrollable weather conditions, and diverse interests of a mixed group of people into a productive hunt requires managing mood, time, and activities as well as "sense-making" of the various elements and steps. To translate driving through the Arctic night and stopping to take pictures in the dark, cold Arctic night into an exciting and successful hunting experience with a price tag requires skilful discourse work: telling stories about the Northern Lights, local places, and historical events; engaging the guests in photographing through instructions and prompts; keeping up a light conversation while driving, waiting and sipping the hot juice by the fire, and gently managing the guests in and out of the car with the tempo of the product.

This is the heavy-lifting part of the hunt for Lucy and Matti. While the co-construction of value in tourism is a well-documented practice in the literature (see e.g. Jensen and Prebensen 2015; Michopoulou et al. 2021), the post-modern version puts additional demands on it. The service provider is expected not only to create affordances for interaction and participation but also to provide real and rare experiences in the most subtle, invisible ways possible, in keeping with the idea of post-modern tourism where the tourists want to experience themselves firsthand (Xiang et al. 2021). For a successful post-modern experience, tourists need to feel that they are experiencing unstaged events and have a genuine connection to the place and it's people. At the same time, the hunt must be managed in terms of time and action: the tourists' safety and comfort need constant care, and they need support for working with the cameras in the cold and darkness. This management needs to be done discretely without any sense of shepherding or hurrying. The tourists pay for excellent service, which translates into a sense of having no time limits, just enjoying themselves while waiting for this natural wonder to emerge.

The discourse work has multiple goals: to amalgamate the different elements in the assemblage together, to connect the product both to the Arctic region and to position Lucy and Matti as the best Aurora hunters in the area, and to intensify the hunt experience. Creating a mood and managing a hunt, which includes a lot of waiting and the possibility of disappointment from not seeing the lights; and transforming this into a happy customer with a feeling of money and time well spent, requires highly complex and skilful discourse work. Lucy and Matti excel at all this but are especially skilful narrators. Their narrative skills are honed by their academic training and previous professions. Lucy recognizes that their ability to tell both visual and oral stories is the winning quality in making a successful Aurora chase experience. Narratives transform and anchor the tourism experience (Mathisen 2019; Kelly-Holmes and Pietikäinen 2014). In their local context, their oral and visual narrating competence is intertwined with native-level English and cosmopolitan trajectory. Together, they produce a cutting difference as smooth, skilful intercultural communicators with an understanding of cultural sensibilities, all fitting well into the international, English-dominated niche market.

Lucy and Matti use narratives for multiple purposes: to hold the various elements and contradictions together—to entertain tourists, and to legitimate themselves. Lucy gives a specific role to her narratives: to foster connections and a sense of sharing. She translates the social and political history of the region into stories, which she weaves in the locations that the hunt takes us: she uses the river to tell about the shifting border between Sweden and Finland, the Finnish bread or potatoes to narrate regional differences among the villages we pass, and the occasional old wooden buildings that survived the Second World War tell about German-Finnish relationships and presence of German soldiers in these villages at the time. She emphasizes the importance of having a story to tell rather than just selling a product or providing an experience. This narration of cultural and historical traditions and processes gives their Aurora product an edge. It fits well within the post-tourism market. Lucy feels that narratives enhance the tour experience by teaching people not only about the Northern Lights but also the environment and social history of the area. This builds into the experience of a successful hunt, making it worthwhile.

Narratives are powerful discursive tools for de/reterritorializing processes. They have the ability to make connections across times, places, bodies, and objects. Moreover, narratives produce affective impulses and intensify the experience by evoking affective resonance through nostalgia, memory, and speculation. In that sense, narratives are ultimately about sharing, creating a sense of temporal connection between the narrator and listener, a communicative bond. They also allow for sharing beyond the immediate moment—the tourists can "renarrativise" this to recount the experience to their friends and family, or through their social media postings (Kelly-Holmes and Pietikäinen 2014). Narratives are an important tool for added value in case of a no-show (Kelly-Holmes 2019): they can help to decenter the disappointment and refocus on the experience of changing: the less spectacular the event, the more narrative labour is needed to restablize aurora assemblage.

Lucy and Matti hit the jackpot with their Aurora program. It quickly became popular among international tourists. Already in the first season, it was profitable enough to keep Lucy and Matti going. They were able to pay themselves a modest salary. This was not close to their income in their previous life, but sufficient for them to consider expanding their business. The experience of the first two seasons has shown them that they need to purchase land in order to secure easy access to the open northern skies. They started looking for a property to buy, and by the end of the year, they purchased a cottage with a lake view. This has proven to be one of their best business decisions ever.

5.4 Aurora Boom: "It Has Been a Fantastic Year for Aurora Chasing, with a Success Rate over 90%"

Fig. 5.4 Light painting while waiting for Auroras

> Before we get too comfortable and tired by the fire and cup of coffee, our guide Matti announces it is time to continue the chase. We return to the car and drive to our next Aurora hunting grounds for a while. We park the van near a frozen river. Now, with some routine, we take our cameras out and head towards our own spots to take pictures. Matti introduces light painting, which involves shining light onto trees, snow hills, and abandon shed, while photographing with a long exposure time, resulting in various lines and circles of light in the photo. I forgot about the Auroras and had fun with light painting. After a while, Matti points to a pale greenish streak in the sky. It is difficult to tell whether this is faint Aurora or light pollution from a nearby town. Matti looks at his Aurora broadcast, which shows high activity in the sky right now, but there are still some clouds. Matti suggests that we take pictures of the light as the camera lens can see the Auroras better than our eyes. We all do this, and after looking at the pictures, we come to the conclusion that the streak is most likely Aurora. This speculation allows us to happily claim that we saw Auroras, even if very faint ones. The hunt has been successful. (Fieldnotes, January 2018)

After two years of development, the Aurora chase has become a highly successful product, turning Lucy and Matti into full-blown entrepreneurs with a solid reputation and income. Lucy feels confident in marketing, with a 97% success rate for their hunts on social media. "If weather conditions are right and we know we can move the guests around, we chase Auroras. There is actually an 85–90% chance of seeing the Northern Lights and finding a gap in the clouds". In the event of a no-show, Lucy and Matti have developed a wide range of techniques to manage the absence of the Northern Lights, including night photography, light painting, offering warm juice and brownies, and telling stories of the region. These activities enable them to end the chase by saying, "We did our best, and we had a great time". In the end, tourists have experienced the thrill of the chase through the dark Arctic night and have pictures to prove it (Fig. 5.4).

The economic success and growing demand for Northern Lights products radically altered the local attitude towards Aurora chasing products. Lucy describes how they have become "legit washed" and are now being promoted by local tourism services. The Aurora chase has transformed from a suspicious "scam" to a popular part of the local tourism industry. Lucy and Matti also face increasing competition from local entrepreneurs offering various versions of the Aurora chase. The accelerating tourism boom and increased competition prompted Lucy and Matti to reassemble their Aurora product again. They make two decisions: they expand their business by developing a new product, *Aurora camp*; and further differentiate themselves from the competitors through a carefully curated social media presence.

The cottage purchase enables the development of the new product, Aurora camp. The cottage with a lake view becomes a game shifter for reassembling their tourism products. It reconfigures both their business model and ties to the local community. It allows Lucy and Matti to diversify their business and recode themselves as locally rooted business owners.

The cottage checks all the boxes for a unique Arctic experience and local legitimacy. It is located next to a popular nature park, beside a tranquil Arctic lake surrounded by fells, with an open view of the Northern skies, yet is easily accessible from tourist resorts, airports, and service centres. Lucy discovers the location is so great that even locals go "Wow how did you get this, oh my god, this is amazing". The property also has an outdoor, sheltered teepee, a *kota*, with a fireplace. Also, the kota is locally validated, and Lucy explains that "even the Finns are happy to go in our kota". The cottage also simplifies and improves the management of their business. Lucy describes how "We no longer have to clean the path, and just knowing that we have the firewood and a flushing toilet means we do not have to prepare; we know it is clean". The cottage becomes their Aurora tourism headquarters.

The cottage is also a big hit internationally. The cottage with iconic Arctic views has a visual currency that has brought international visibility to Lucy's and Matti's company. It has served as a photo shooting location for an international feature article about Finland being the happiest country in the world, as well as for a fashion piece for a global fashion magazine. In addition, many international influencers and television crews use the cottage to film Arctic-themed stories. This global visibility, with high symbolic value, has helped put them on the Arctic destination map for the international market. Lucy thinks the cottage is perfect, as it has a feeling of remoteness but in a wholesome, friendly way.

Locally, the cottage recodes them as landowners in one of the most iconic Finnish Lapland regions, rooting them in the local community in a novel way. In the local political economy, the purchase is an index of a successful business and a sign of a decision to root themselves for a more permanent stay. Lucy notes, "The property has given us status and a sense of permanence that we did not have before". Buying property is part of the process of becoming more local and more serious business people.

The cottage makes the new product, Aurora camp, both possible and necessary. The financial investment in the cottage triggers a familiar chain reaction of business growth through product diversification and employee hiring. It restructures their business. However, despite the potential of the cottage, reassembling the original Aurora chase into two products has been a challenge.

To make the most of the cottage, they split the Aurora chase into two products: the high-end, "sophisticated and professional" Aurora chase and the "quick and easy" Aurora base camp-product designed for the bulk market. Lucy explains that the base camp refers to "the easy, comfortable and safe experience", being in one place (their cottage) rather than "chasing" after the lights with the car across the Arctic terrain, which is the key activity for their main Aurora chase program.

By splitting the original Aurora chase into two products, the cottage can offer different levels of experience and cater to different budgets. The high-end Aurora chase provides a more immersive experience requiring a longer stay, while the Aurora base camp is a shorter, more budget-friendly option that still allows customers to see the lights. The camp product provides easy access to northern skies without light or noise pollution simply by stepping outside of the warmth and comfort of the *kota* with an open fire and refreshments. They locate this product in the early evening at a lower price to accommodate families with kids and other customers who want to do something in the early evening.

At first, they struggled to make the Aurora camp product work. "It is a different beast", Lucy bemoans. While the material conditions of the cottage and *kota* were perfect for the camp, the discursive and affective aspects needed reworking. Lucy initially tried to reterrotorialize the affective and discursive tools, especially the narratives from the Aurora chase to the Aurora camp, but she quickly realized that they do not work in this new terrain. At the camp, the tourists were in tour mode and had no interest in engaging with Lucy or her stories. "It is a different level of intimacy and connection, no matter how hard I try", Lucy summarizes. The tourists were not prepared to listen, leading to a lack of connection to the place and a different perception of the Arctic environment. They had a few complaints, which was a clear sign to Lucy and Matti that the expectations and offer did not meet. Lucy explained: "I am trying, and I am skilled in engaging with people, but still, afterwards, I am broken, after spending 3 hours on a camp with 15 people, who say 'is that all there is?', and I can only say 'do you want a marshmallow?".

Only when Lucy changes the discursive and affective dimensions of the Aurora camp assemblage does it start to function as a successful tourism product. Lucy starts to organize the Aurora camp around chilling in comfort, with family or a group of friends, suitable for everyone of all ages. It becomes a "lazy version of the chase": spending a camp night out chilling out together, with the possibility of staying in the warmth of the open fire at *kota*, popping out just a few meters for Aurora gazing or feeling the Arctic night. Instead of hunting, they code the camp with relaxing, unwinding, and spending quality time together. They gently encourage tourists to take pictures with their own cameras, as a camera can change their feeling of "wasting my time". With these changes, the Aurora base camp product gains a profile as an alternative way to spend the early Arctic evening together. It has become popular, especially among families.

To run these two products profitably, material expansion is required and yet again the Aurora assemblage needs to be reassembled. This entails hiring an employee and buying a second van. Also the tempo accelerated. They get really busy: they could have over thirty customers on their busiest nights, split between Aurora chases and Aurora base camp.

The tension between the local, nation-state-oriented tourism assemblage and the post-modern, global tourism assemblage of Lucy's and

Matti's business persists. As Lucy and Matti's business grows, the differences in the operation logics become more pronounced. While most local tourism operators strive for big numbers, Lucy and Matti prioritize quality for a relatively small group of people. Their profits come from quality, not quantity. Due to this underlying difference, Lucy thinks the locals do not get their marketing strategy around lifestyle and multisensory experience. To navigate the local political economy, Lucy uses a range of codes for their business. She prefers to code them professionally through their academic background in business encounters: "We will carry the academic banner, and I think we will keep that specialty line".

Lucy's multiple trajectories, practices, and skills continue to be considered *nomadic* by locals. They are not easily coded with the fixed local categories but rather render more cosmopolitan flows of mobility, multiplicity, and change. Locally, she has changed from being coded as "a foreigner" to belonging to a local "exotic club", a category that refers to a non-Finnish historical trajectory but permanent residency in the region. Lucy explains that locals identify her as a foreigner living in the village, and she is fine with that, even though she fits uncomfortably into this category. Lucy has noticed that locals assume she cleans houses and that she came to Finland to improve her financial situation. But this is not the case. She is a mobile, migrant mother with a university degree and a prospering business. Her codes are dynamic and rhizomatic, cutting across and combining multiple subject positions. She exemplifies nomadic assemblage.

Nomadic assemblage functions along with nonhierarchical and fluid organization, entering new combinations without arbitrary hierarchical limits (Nail 2017: 380–87). With Lucy, this translates into difficulties in getting across her ideas to the local tourism establishment. Her out-of-the-box ideas and ways of working make her locally controversial. She has realized that this is interpreted locally as a lack of understanding of local conditions or insufficient support for local efforts and practices. Her behaviour is subject to public scrutiny. She has tried to reterritorlize her post-modern tourism practices into local, nation-state tourism assemblage but in vain. She thinks that her ideas do not fly easily in the local business circuits partly because of their novelty but also because she lacks sufficient Finnish skills to simplify the ideas to be understood. She has tried to overcome the linguistic barrier by sitting next to local business people at dinners, sending follow-up emails, and inviting them to lunch, but this, too, has been in vain. Eventually, she finds a connecting line with food. She is an excellent cook, and she begins to recode lunches and dinners that she organizes for local partners and business people in their cottage with commensality. This is the start of her own local professional network and emerging wild food assemblage: "Food is a bridge", Lucy concludes.

5.5 Curating and Coding the Virtual Aurorascape

Fig. 5.5 Instragrammable moment

"Arctic, Lapland Finland, Scandinavia, Best Aurora chase, Double Northern Lights hunt, fox fire, lake, autumn winter, snow, night, the Big Dipper, escape, photography, reflection, portraits, light painting, private, colour, holiday". These are some of the keywords and hashtags that Lucy uses in their three sentence long Instagram caption. They are carefully selected based on Lucy's understanding of their market and designed to evoke affective connections to the region and their Aurora tourism products. She uses them to frame and connect their products to search engines so that people can find them when they are doing online searches (Fig. 5.5).

To dip into the virtual flow of creating pre- and post-visit tourism experiences, a vital part of post-tourism (Jansson 2018), Lucy and Matti began to develop their social media presence. In comparison with traditional tourism, post-tourism value creation includes several shifts: from value chain to value network, from a physical environment to virtual, from a product to capacity, from static and fixed to emergent and shapable, from linear and sequential to simultaneous (Xiang et al. 2021). In assemblage terms, virtual post-tourism value creation becomes a rhizome of simultaneous possibilities, requiring multiple skills and constant assembling work from Lucy and Matti. The two core skills are curation and coding.

For their *virtual Aurorascape*, Lucy curates intensity. Curation refers to an act of selecting, organizing, and presenting a collection of items in a meaningful way. It is typically used in museums and art exhibitions but is also increasingly used in social media content production (Davis 2017; Richards 2024). Curation also plays a central role in designing tourism destinations. Lucy combines these curation elements to create a virtual Aurorascape. This is right up her alley, as she was a professional curator in her previous life. She applies her expertise in curating and knowledge about her markets and products to shape their social media into an Aurorascape that conveys a specific message.

Their website and Instagram page become carefully curated into a distinctive virtual Auroraspce, enabling travellers to speculate about future hunts or relive past ones. As such, it also became a trendsetter among the local Aurora entrepreneurs, who follow her Instagram page and webpages with a keen eye, resulting, at times, in copying. Lucy describes how she speaks the same language visually and linguistically as their primary market area, which consists of English-dominant countries.

The virtual Aurorascape is strategically recoded to engage their target, niche international market and to differentiate their products from competitors. Lucy acknowledges their excellent capacity and skills to code their message both visually and textually. They design a distinct visual language using their professional skills and training in fine arts and photography. Lucy identifies the photos' high quality and traditional composition as essential to their visual language. They also use photos very selectively and strategically: "We do not want our feed to be polluted by pictures of guests wearing overalls under the Northern Lights as this may give an impression of coldness. Rather, we aim for images that tell the main story: seeing the Northern Lights with a touch of other nice activities, like putting the fire on, sitting by the fire, playing with a sparkler, and having a good time. We also look for pictures with faces and camera equipment". Featuring cameras allows them to index one of their key codes for distinction: they are professional and provide high-quality equipment. This professionalism was also underlined by using the title Dr. for Matti and mentioning his photo exhibitions in bylines.

Lucy is a wordsmith. She codes for affect and connection. Lucy strategically varies language to manage the perception and reach of the product: She shifts the product's focus by varying the use of verbs such as "emerge yourself" versus "discover" to appeal to different audiences. When she targets the corporate incentive market she boosts concepts such as teamwork, engagement, and potential, and gets far more hits than she would for using "emerge" or "discovery". The codes for family markets include "sharing" and "enjoying". She designs the text to appeal to all senses, using expressions like "listen to silence" or "hear the sound of snow under your feet". She also avoids certain codes, one of them being "luxury", which she describes as a language marker: luxury means a lot of work and high expectations, which would be impossible to meet as people's expectations of luxury vary across cultures and ages (Iloranta and Komppula 2022; Thurlow and Jaworski 2012; for wordsmith see also Duchêne 2019; Thurlow 2019).

The virtual Aurorascape also unfolds for what Lucy calls "gentle marketing". Resonating with the post-tourism frame, Lucy knows that their market does not respond well to hard marketing, so they need to be delicate in their approach. Lucy creates virtual connections in the photos through hashtags, tagging, and keywords. This creates an emergent and evolving molecular rhizome of Aurora connections.

Social media practices also influence how the Aurora products are created (see e.g. Gershon and Prentice 2021; Kelly-Holmes 2019; Lim and Childs 2020; Pak and Hiramoto 2022; Smith 2018; Thurlow and Jaworski 2012). The demand for visual consumption and distribution of the Arctic via social media is so high that Lucy describes part of their work as creating "Instagrammable moments": visually delicious affordances that render easily to selfies, unique to the hunt yet repetitive enough to fit into the Instagram genre and marketing. As a result, some locations and activities in their Aurora cottage have become visual Aurora hotspots. These include an open fire, chopping wood and putting it in a fireplace in the kota, the lake view, snow cones, and things covered by snow. These hotspots meet Instagram's visual criteria of being photogenic, artistic, and aesthetic (Conti and Lexhagen 2020; Smith 2021). They are interactive and attractive, with the potential to impress viewers and gain recognition on Instagram. There is a certain recirculation and repetition effect in the visual consumption of the Arctic on social media; a particular photo or video can start to trend because a famous influencer takes it or it is otherwise attracting much attention, with a result, the next visitors want to do similar kind of picture, too.

The publicly accessible website and Instagram feed bring challenges, too. After carefully designing their products' visual language and textual descriptions, their style gets "picked up by other companies". After some initial irritation, Lucy takes it as a compliment and, in exchange, keeps an eye on their competitors' social media outputs. She is not too worried about the competition as she believes that both their message and products are quite complicated, as it is less about "putting the product into the sellable platform but rather about the ability to change per audience and per season". However, to have some control over their social media messages, they begin to trademark their product descriptions with their logo. This legal recoding marks and protects the descriptions as their property, preventing them from being copied or altered as they circulate.

The virtual Aurorascape becomes a direct link between them and their customers. Over half of their customers make bookings directly via their website rather than using travel agencies or local tourism information services. This is yet another aspect of Lucy's and Matti's post-modern tourism infrastructure: they can do business directly with their clients without local and national facilitators or gatekeepers.

5.6 Busted by Pandemic: From Pivoting to Bubble Tourism

Fig. 5.6 An Arctic Airport during the COVID-19 Pandemic

"I dread Mondays nowadays", Lucy says in the middle of the tourism bust caused by the global COVID-19 pandemic. The pandemic emptied the Finnish Arctic of tourists after a five-year boom. The change was dramatic. Previously thriving ski resorts became ghost towns. For Lucy and Matti, the travel restrictions, together with the social distancing policy meant that the niche Aurora market of international tourists vanished almost in the blink of an eye. Lucy's calendar is suddenly empty, and she finds herself "busy pretending to be busy".

When the Finnish government implemented isolation policies and travel restrictions in March 2020, it effectively stopped all international tourism in the Arctic (Fig. 5.6). Lucy and Matti were out of business like everyone else in the tourism industry. Without international travel, their market base is gone, and they must figure out how to make a living under uncertain pandemic conditions. Like many other countries, Finland went into lockdown: schools and public institutions were closed, remote working was mandated, and at most ten people could participate in a public gathering.

Lucy and Matti spent their lock-out time with their children at their cottage, where they were able, like all Finns, to go out for walks, fish, and pick berries. Lucy estimates they will survive for at least a year as they have had a few successful seasons, and now their strategy is to "just wait and see". As the pandemic was prolonged with travel restrictions and social distancing measures, it became clear that the international tourism season of 2020–2021 would be lost. Lucy recalls how only a year ago, they were fully booked for December, and now there is not a single booking.

The sudden disappearance of the tourists makes the Arctic tourism assemblage unfunctional: no bodies, no desire, no activities, no money. It also made Lucy's and Matti's Aurora products unfunctional: without international tourists, their carefully crafted re/deterritorialization narratives and coding for genuine Arctic night experience no longer worked. Also, they could no longer code their work and skills with global, cosmopolitan, post-modern tourism flow, as it disappeared, too. Instead, the national Lapland tourism assemblage made a comeback and became the dominant organizing politics of the local tourism industry.

This shift in the politics of tourism assemblage left Lucy and Matti on the outskirts of the local tourism terrain. In this moment of crisis, the typical national indexes of belonging became foregrounded: Finnish skills, local roots, and tourism traditions. The tourism space becomes stratified with nation-state logic, leaving little smooth space for Lucy and Matti to carve their belongingness. In Deleuzian-Guattarian terms (1980/1987), this movement marks a shift from smooth space to stratified space: the smooth space is yet unmapped, creative, and emerging (like a sea), while the stratified space has norms, rules, and grids (like coordinates on a sea). In stratified tourism space, the categories and boundaries become inflexible. "As a foreigner, it is hard to create a space for you", Lucy notes. Under the new conditions, it became increasingly difficult for them to reterritorilize their trajectories, skills, and business in the traditional tourism terrain, as well as to recode their tourism business to the national tourism flow. Lucy participated in some online tourism community meetings and events, but it quickly became clear to her that "they did not fit in". Ideologically, at this point, their cosmopolitan trajectories and international skills did not fit into the national and local tourism histories and infrastructures. Lucy also feels sidelined due to the limits of her Finnish skills in new virtual environment of meeting platforms. She finds that the lack of physical cues used by Finns makes communication more restrictive for her, and she misses face-to-face meetings where she could "mildly follow and mildly participate" in conversations. She becomes muted.

Under the new conditions, a division emerges between "the old school," established tourism companies, organized by the logic of state assemblage, and the new "lifestyle", more nomadic companies, which are positioned as economically less serious and therefore less deserving of financial support than the established companies which have their decades, if not generations of work at stake. Lucy rejects both categories. She works hard to get across to the establishment that their business is not a light lifestyle company but a hardcore business with successful numbers to show it. But she does not get through to them. She describes the lifestyle company as part of their marketing strategy, a version of the Goliath story, but she hits a brick wall: "They just do not get it", Lucy summarizes. However, for economic survival, they need to keep connected with the local business community and find a way to engage with the current national tourism assemblage. Lucy feels that they cannot be seen as "an outsider company". Thus, they began to take part in the various local pivoting attempts.

Under the COVID-19 conditions, pivoting becomes a local index of "team spirit" and "commitment to working for the common goal", as Lucy describes it. The idea is for entrepreneurs to modify and transform their products into "Covid friendly products" to fit into the pandemic landscape (cf. McDonald and Gao 2019; Wilson-Mah and MacRae 2024). This transformation is further supported by the Finnish government which issued special COVID-19 funding for the tourism industry nationwide to help them to adjust to the COVID-19 condition. Pivoting is a particular kind of discursive de/reterritorializing process: an attempt to move tourism products from the flexible, global tourism markets into the specific terrain of pandemic tourism for national tourists with traveling restrictions and social distancing.

Lucy and Matti engage with the new COVID-19 reality, pivoting with apprehension. At first, they try various strategies to adjust to the changed conditions, such as lowering prices, changing the maximum number of participants, and offering private tours. But nothing helps. Lucy was adamant that there was no business for them as their product was "fundamentally not suited for domestic markets" as Finns do not want to have advice on seeing the Northern Lights, let alone pay for it. As there is no desire, there is no possibility to design a functional assemblage. Lucy also finds the development of COVID-19-friendly products pointless in their case because "the Northern Lights will always be locational", and the chases work only for international markets. Lucy believes that "Auroras will be our saving grace" in the long run. They kept the Aurora products intact and shelved them, waiting for the post-pandemic times.

At this point, pivoting becomes a strategy for both economic survival and communality. Lucy began planning COVID-friendly products around a sense of connection, especially with nature and people. This attempt was supported by government funding, which Lucy and Matti applied for and received. Initially, they hesitated to participate in this kind of project economy (see Chapter 3) as they feel it creates a false economy and is probably not worth it. Lucy feels that such a project application would steer their business in a direction she is unsure they want to go or that makes sense financially. After receiving the funding, Lucy racks her brain to redesign their products, discourses, and market. She is considering moving towards digital products, e.g., cooking workshops online and sharing a digital meal together. Another idea is to make a beautiful coffee table book about Arctic wild food.

After a few months of pivoting, she stops. She feels that there simply is just too much uncertainty and unpredictability in the world to be able to plan ahead, create strategies, or envision the future. "It feels like we are in a fog, unable to see anything, no horizon, no anchoring points". It has become impossible to know what the next best move is. She thinks "fuck it", breaks free, and finds her nomadic line of light: she will use the time to develop, network, and rethink.

As the pandemic was further prolonged, they discovered their COVID-19-friendly product: Airbnb in nature, an Arctic version of bubble tourism for the national market. Bubble tourism involves creating a controlled and isolated environment for visitors, where they can engage in recreational activities while minimizing the risk of exposure to external factors, such as infectious diseases or security threats (Fusté-Forné and Michael 2023). Many Finns who lived in cities sought temporary relief and geographical expansion of their bubbles by renting out cottages in remote locations, facilitated by remote schooling and working practices at the time. Lucy's and Matti's cottage was perfect: it was big enough to facilitate a bigger group, still allowing each member some private space, something that people were craving after cramped pandemic living conditions in the cities. In addition, there was lots to do in the surrounding area. The isolated location of the cottage, far away from the contentious resort area but close enough to travel routes, made it a pandemic haven for Finnish tourists. Under the exceptional conditions created by the global Pandemic, Lucy and Matti are able to combine the logics of territorial and state assemblage in creating a bubble tourism product in the isolation of their cottage in the Arctic.

Lucy and Matti became Airbnb entrepreneurs. The cottage was rented out through the pandemic, and this income provided an economic lifeline for them, covering their daily costs and mortgage.

They reterritorilized their cottage from international, post-tourism Aurora tourism to domestic tourism bubble markets. They recode the cottage, emphasizing the nearness of the nature park and its spaciousness. Lucy caters to Finns' desire for a good bargain by modifying the pictures and description of the cottage to create a positive surprise for the tourists when they see the beauty of the cottage.

The Airbnb business is time-consuming and hard work: accommodating all wishes, responding to the constant flow of messages, and above all, constant cleaning. The Airbnb business turns Matti and Lucy into "basically cleaners". Lucy jokes about cleaning being a big equalizer, as many other local entrepreneurs are also cleaning their premises. Lucy continues to believe in their Aurora products and feels that they simply need to wait for international, high-end tourists to return. Meanwhile, they will economically get by renting the cottage. They wait.

5.7 Back to Business: "Best Year Ever!"

Fig. 5.7 A busy Arctic airport

Aurora tourism quickly picked up as the pandemic travel restrictions eased up, along with the possibility of international bubble travel. Lucy and Matti resume working on their Aurora products. They stop renting their cottage as they found their perfect COVID-19 Aurora product for international markets: a private Aurora chase. It was ideal in several ways. It provided the safety bubble required for the pandemic's continuing social distancing measures, and it was easy to manage between Lucy and Matti as the tourists knew each other already. The cottage is once again used as a base for their chase. They have become super-busy, even busier than before the COVID-19 pandemic (Fig. 5.7).

Lucy was right all along about their Aurora products: they are perfect for the international markets, and they only had to wait. Others have noticed this as well: Lucy counted that over forty Northern Lights products are now marketed in the winter resort in which they operate. Lucy has also carved out a place for herself in the local business community. For any challenges regarding her position or legitimacy locally, she answers now by saying "I am local with different skills and trajectory".

After the travel bans were lifted in 2022, Lucy and Matti travelled abroad with their children to visit their families and friends after a long break. Lucy took the time to contemplate their future dreams and directions for their business and life in general. When they returned to Finland, she sent me a picture of the kitten they had gotten. In 2023, they purchased more land and a new cottage.

References

Anderson, Ben. 2009. 'Affective Atmospheres'. *Emotion, Space and Society* 2 (2): 77–81. https://doi.org/10.1016/j.emospa.2009.08.005.

Bertella, Giovanna. 2013. 'Northern Lights Chase Toursc: Experiences from Northern Norway'. *Journal of Northern Studies* 7 (2): 95–116.

Brekke, Asgeir, and Alv Egeland. 2012. *The Northern Light: From Mythology to Space Research.* Berlin: Springer.

Conti, Eugenio, and Maria Lexhagen. 2020. 'Instagramming Nature-Based Tourism Experiences: A Netnographic Study of Online Photography and Value Creation'. *Tourism Management Perspectives* 34 (April). https://doi.org/10.1016/j.tmp.2020.100650.

Davis, Jenny L. 2017. 'Curation: A Theoretical Treatment'. *Information, Communication & Society* 20 (5): 770–83. https://doi.org/10.1080/1369118X.2016.1203972.

Deleuze, Gilles, and Felix Guattari. 1980/1987. *A Thousand Plateaus: Capitalism and Schizophenia.* Translated by Brian Massuni. London: Bloomsbury Academic.

Duchêne, Alexandre. 2019. 'Unequal Language Work(Ers) in the Business of Words'. In *The Business of Words,* edited by Crispin Thurlow, 21–36. London: Routledge.

Edensor, Tim. 2013. 'Reconnecting with Darkness: Gloomy Landscapes, Lightless Places'. *Social & Cultural Geography* 14 (4): 446–65. https://doi.org/10.1080/14649365.2013.790992.

'Explore Finnish Lapland—Europe's Last Great Wilderness Visit Finland'. n.d. Accessed 1 January 2024. https://www.visitfinland.com/en/places-to-go/lapland/.

Falck-Ytter, Harald, Torbjörn Lövgren, and Harald Falck-Ytter. 1999. *Aurora: The Northern Lights in Mythology, History and Science*. Hudson, NY: Bell Pond Bks.

Fusté-Forné, Francesc, and Noela Michael. 2023. 'Limited Tourism: Travel Bubbles for a Sustainable Future'. *Journal of Sustainable Tourism* 31 (1): 73–90. https://doi.org/10.1080/09669582.2021.1954654.

Gershon, Ilana, and Michael M. Prentice. 2021. 'Genres in New Economies of Language'. *International Journal of the Sociology of Language* 2021 (267–268): 117–24. https://doi.org/10.1515/ijsl-2020-0055.

Gonçalves, Kellie. 2020. *Labour Policies, Language Use and the 'New' Economy: The Case of Adventure Tourism*. Cham: Palgrave Macmillan.

Harvey, David. 1989. 'From Managerialism to Entrepreneurialism: The Transformation in Urban Governance in Late Capitalism'. *Geografiska Annaler: Series B, Human Geography* 71 (1): 3–17. https://doi.org/10.1080/04353684.1989.11879583.

Hautajärvi, Harri. 2014. *Autiotuvista lomakaupunkeihin - Lapin matkailun arkkitehtuurihistoria*. Aalto University. https://aaltodoc.aalto.fi/handle/123456789/19766.

Heimtun, Bente, and Bente Haug. 2022. 'The Development of the Northern Lights Tourism Network'. *Annals of Tourism Research Empirical Insights* 3 (1): 100031. https://doi.org/10.1016/j.annale.2021.100031.

Heimtun, Bente, and Brent Lovelock. 2017. 'Communicating Paradox: Uncertainty and the Northern Lights'. *Tourism Management* 61 (August): 63–69. https://doi.org/10.1016/j.tourman.2017.01.017.

Heimtun, Bente, Gunnar Þór Jóhannesson, and Seija Tuulentie. 2015. 'Northern Lights Tourism in Iceland, Norway and Finland'. *Septentrio Reports*, no. 1 (February). https://doi.org/10.7557/7.3266.

Helander-Renvall, Elina. 2006. *Silde: saamelaisten myyttejä ja tarinoita*. Utsjoki: Elina Helander-Renvall.

Iloranta, Riina, and Raija Komppula. 2022. 'Service Providers' Perspective on the Luxury Tourist Experience as a Product'. *Scandinavian Journal of Hospitality and Tourism* 22 (1): 39–57. https://doi.org/10.1080/15022250.2021.1946845.

Ingold, Tim. 1993. 'The Temporality of the Landscape'. *World Archaeology* 25 (2): 152–74. https://doi.org/10.1080/00438243.1993.9980235.

———. 2021. *The Perception of the Environment: Essays on Livelihood, Dwelling and Skill*. 1st ed. London: Routledge. https://doi.org/10.4324/9781003196662.

Jansson, André. 2018. 'Rethinking Post-Tourism in the Age of Social Media'. *Annals of Tourism Research* 69 (March): 101–10. https://doi.org/10.1016/j.annals.2018.01.005.

Jensen, Øystein, and Nina Prebensen. 2015. 'Innovation and Value Creation in Experience-Based Tourism'. *Scandinavian Journal of Hospitality and Tourism* 15 (sup1): 1–8. https://doi.org/10.1080/15022250.2015.1066093.

Jóhannesson, Gunnar Thór, and Katrín Anna Lund. 2017. 'Aurora Borealis: Choreographies of Darkness and Light'. *Annals of Tourism Research* 63 (March): 183–90. https://doi.org/10.1016/j.annals.2017.02.001.

Katila, Saija, Ari Kuismin, and Anu Valtonen. 2023. 'Natural Light as Affective Force in Organizing Practices'. In *A Research Agenda for Organization Studies, Feminisms and New Materialisms*, edited by Marta B. Calás and Linda Smircich, 55–72. Cheltenham: Edward Elgar Publishing. https://doi.org/10.4337/9781800881273.00008.

Kelly-Holmes, Helen. 2019. 'The Linguistic Business of Marketing'. In *The Business of Words*, edited by Crispin Thurlow, 36–50. London: Routledge.

Kelly-Holmes, Helen, and Sari Pietikäinen. 2014. 'Commodifying Sámi Culture in an Indigenous Tourism Site'. *Journal of Sociolinguistics* 18 (4): 518–38. https://doi.org/10.1111/josl.12092.

Lim, Heejin, and Michelle Childs. 2020. 'Visual Storytelling on Instagram: Branded Photo Narrative and the Role of Telepresence'. *Journal of Research in Interactive Marketing* 14 (1): 33–50. https://doi.org/10.1108/JRIM-09-2018-0115.

Mathisen, Line. 2019. 'Storytelling: A Way for Winter Adventure Guides to Manage Emotional Labour'. *Scandinavian Journal of Hospitality and Tourism* 19 (1): 66–81. https://doi.org/10.1080/15022250.2017.1411827.

McCormack, Derek P. 2008. 'Engineering Affective Atmospheres on the Moving Geographies of the 1897 Andrée Expedition'. *Cultural Geographies* 15 (4): 413–30. https://doi.org/10.1177/1474474008094314.

McDonald, Rory, and Cheng Gao. 2019. 'Pivoting Isn't Enough? Managing Strategic Reorientation in New Ventures'. *Organization Science* 30 (6): 1289–1318. https://doi.org/10.1287/orsc.2019.1287.

Michopoulou, Eleni, Nikolaos Pappas, and Elena Cavagnaro, eds. 2021. *Revisiting Value Co-Creation and Co-Destruction in Tourism*. London: Routledge. https://doi.org/10.4324/9781003245537.

Milani, Tommaso, Erez Levon, and Ruth Glocer. 2019. 'Crossing Boundaries: Visceral Landscapes of Israeli Nationalism'. *Sociolinguistic Studies* 13 (1): 37–56. https://doi.org/10.1558/sols.36209.

Mortensen, Ellen. 2017. 'Mood, Method and Affect: Current Shifts in Feminist Theory'. *Feminist Encounters: A Journal of Critical Studies in Culture and Politics* 1 (October): 1–9. https://doi.org/10.20897/femenc.201702.

Nail, Thomas. 2017. 'What Is an Assemblage?' *SubStance* 46 (1): 21–37.

Pak, Vincent, and Mie Hiramoto. 2022. '"Itching to Make an Impact": Constructing the Mobile Singaporean Voluntourist in Instagram Travel Narratives'. *Social Semiotics* 32 (3): 332–54. https://doi.org/10.1080/10350330.2020.1766263.

Pietikäinen, Sari, Helen Kelly-Holmes, Alexandra M. Jaffe, and Nikolas Coupland. 2016. *Sociolinguistics from the Periphery: Small Languages in New Circumstances*. Cambridge: Cambridge University Press.

Ren, Carina, and Mette Simonsen Abildgaard. 2021. 'Greenlandic Independence and Tourism Futures: Exploring Modern and Ethnic Logics'. In *Tourism, Climate Change and a Geopolitics of Arctic Development*, edited by Derek Hall. CABI Publishing. https://www.cabi.org/bookshop/book/9781789246728/.

Richards, Greg. 2024. 'The Curatorial Turn in Tourism and Hospitality'. *International Journal of Contemporary Hospitality Management* 36 (13): 19–37. https://doi.org/10.1108/IJCHM-06-2023-0905.

Rinne, Jenni. 2021. 'Affective Place Experiences and Landscape Meanings among Estonian Native Faith Practitioners'. *Ethnologia Scandinavica* 2021 (51): 61–80.

Smith, Sean P. 2018. 'Instagram Abroad: Performance, Consumption and Colonial Narrative in Tourism'. *Postcolonial Studies* 21 (2): 172–91. https://doi.org/10.1080/13688790.2018.1461173.

———. 2021. 'Landscapes for "Likes": Capitalizing on Travel with Instagram'. *Social Semiotics* 31 (4): 604–24. https://doi.org/10.1080/10350330.2019.1664579.

Snyder, Benjamin. 2016. *The Disrupted Workplace: Time and the Moral Order of Flexible Capitalism*. Oxford: Oxford University Press.

Stroud, Christopher, Amiena Peck, and Quentin Williams. 2019. 'Introduction: Visceral Landscapes (the Inside Story)'. *Sociolinguistic Studies* 13 (1): 7–14. https://doi.org/10.1558/sols.36224.

Sun, Jiaojiao, and Xingyang Lv. 2021. 'Feeling Dark, Seeing Dark: Mind–Body in Dark Tourism'. *Annals of Tourism Research* 86 (January): 103087. https://doi.org/10.1016/j.annals.2020.103087.

Thurlow, Crispin. 2019. 'The (Grubby) Business of Words: What "George Clooney" Tells Us'. In *The Business of Words*, edited by Crispin Thurlow, 1–20. London: Routledge.

Thurlow, Crispin, and Adam Jaworski. 2012. 'Elite Mobilities: The Semiotic Landscapes of Luxury and Privilege'. *Social Semiotics* 22 (4): 487–516. https://doi.org/10.1080/10350330.2012.721592.

Tucker, Hazel, and Eric J. Shelton. 2018. 'Tourism, Mood and Affect: Narratives of Loss and Hope'. *Annals of Tourism Research* 70 (May): 66–75. https://doi.org/10.1016/j.annals.2018.03.001.

Vandenberghe, Frédéric. 2008. 'Deleuzian Capitalism'. *Philosophy & Social Criticism* 34 (8): 877–903. https://doi.org/10.1177/0191453708095696.

Weaver, David. 2011. 'Celestial Ecotourism: New Horizons in Nature-Based Tourism'. *Journal of Ecotourism* 10 (1): 38–45. https://doi.org/10.1080/14724040903576116.

Wilson-Mah, Rebecca, and Katherine MacRae. 2024. 'Pivoting in the Tourism Sector: COVID-19'. In *Tourism and Hospitality Management in Practice: A Case Study Collection*, edited by R. Wilson-Mah, 25–32. London: Routledge. https://doi.org/10.4324/9781003280040.

Wylie, John. 2007. *Landscape*. London: Routledge.

Xiang, Zheng, Jason Stienmetz, and Daniel R. Fesenmaier. 2021. 'Smart Tourism Design: Launching the Annals of Tourism Research Curated Collection on Designing Tourism Places'. *Annals of Tourism Research* 86 (January): 103154. https://doi.org/10.1016/j.annals.2021.103154.

6

Arctic Reconfigurations: From Frozen Representations to Flowing Production

6.1 Machinic Cold Rush

Fig. 6.1 Accelerating Cold Rush machine

The Cold Rush in the Finnish Arctic keeps on accelerating. The combination of the accelerating global warming, changes in the geopolitical landscape, and record-breaking numbers of tourists makes the Arctic

S. Pietikäinen, *Cold Rush*, Language and Globalization,
https://doi.org/10.1007/978-3-031-63995-1_6

region an economic, ecological, and political hotspot. Constant revision of strategies and alliances has become the norm, with assembling becoming a key skill (Fig. 6.1).

As we have seen in this book, the rush for Arctic natural resources has a long history in the region. The first rush for iron ore and copper began in the seventeenth century when the first mines were established in the region. This marked the installation of capitalist and colonial logics, which gradually expanded and deterritorialized indigenous Sámi people as nomads and their traditional lands as extraction frontiers while reterritorializing Finns as settlers. Under these early capitalist and colonial conditions, this transformation was coded as progress and continues to figure in the local decision-making regarding natural resources.

The mining assemblage continues to reterritorialize the idea of progress to new conditions. Originally, it referred to "civilization" in terms of state governance and church's regimes; by the turn of the twentieth century, progress was translated into rights and better living conditions for workers. In times of political unrest and war, during the first half of the twentieth century, the mining assemblage connected the political movement of workers and the communist ideology to the discourse of progress. This was a strong alliance that still continues to produce the Finnish Arctic region as the politically "the most red" in the country. In the latter part of the twentieth century, the mining boom inserted new elements into the assemblage: a growing middle class, global investors, and environmental risks. Currently, mining divides the local communities into pro-mine and anti-mine camps with opposing views on the alliance between mines and progress: for the former, mining continues to bring progress in terms of jobs, people, and income, whereas the latter sees mining as an obstacle for sustainable, green progress. The different mining politics organize mining assemblage differently with different outcome.

While the mining assemblages continues to function, the revaluing attempts of the wild Arctic berries highlight how the assembling work does not always succeed. Despite carefully selecting material and discursive elements and organizing them with a particular logic, the desired outcome may not emerge. Working through a municipal economic development project, the state berry assemblage aimed to create local job opportunities through sustainable foraging practices. The reterritorialization of the berries and foraging from the commons to commodities and jobs did not work for two main reasons: the locals valued their berries more than the exchange value available, and there were not enough suitable bodies in the sparsely populated Arctic to meet the criteria of sustainable foraging. The collaboration between the state and the capitalist assemblage politics did not take root in the local berry patches, which remain territorialized by traditional logics of reciprocal commons. Only the economic conversion of the ordinary Arctic berries into superberries through discursive recoding functioned, as the nutritional materiality of the berries can be translated into a sellable product for the global health markets.

Unlike the mining and berry assemblages, the Aurora assemblage bypasses the state assemblage politics and connects with the territorial and capitalist assemblage politics, resulting in highly functional Aurora chase products. Auroras are firmly rooted in the Arctic territory, and the embodied Aurora experience is only possible in the Aurora zone during the dark winter nights. Revaluing this nature experience with excellent service and a price tag did not make sense locally but made perfect sense to the high-end global, post-modern tourists. However, to navigate the local political tourism economy, on the one hand, and to adapt to changing weather conditions, on the other, the Aurora assemblage requires constant, intensive, and skillful discourse work, such as curating and pivoting. All of this work came to a halt when the global pandemic froze international travelling. Under the radically changed conditions, especially the lack of tourists, the Aurora assemblage could not function any longer. Like in mining and berry assemblages, the Aurora assemblage also needs suitable bodies to work. Indeed, what all these assemblages seem to share is the dependency of the "right" bodies to function. In the sparsely populated Arctic, findings these bodies, be they miners, berry pickers or tourists, requires constant work.

Machinic Cold Rush emerges when we connect the mining, the berries, and the Aurora assemblages. For Deleuze and Guattari (1980/1987), a machine refers to an organized system of assemblages that operates according to certain logics and functions. The machine generates and circulates flows of desire, energy, and logics. The Cold Rush machine produces ongoing revaluing of natural resources by attempting to inscribe functions for them. The main politics for organizing the Arctic natural resource assemblage is capitalist, organizing the assemblage so that it would reterritorialize and recode resources and relations around them in terms of exchange value. The state politics is a close ally of capitalist politics in its desire to organize assemblages so they would produce local jobs and tax revenue and keep the periphery functional and inhabited. In the context of the Cold Rush, the territorially organized assemblages, such as berry commons, provide a counter power to capitalist and state politics. Territorial assemblages align with nomadic thinking in resisting the totalizing capitalist and state politics.

However, the assemblage effects are never totalizing nor homogeneous. In the context of the Cold Rush, the power dynamics of the capitalist and state alliance are resisted and bypassed by nomadic and territorial local bodies, berries, and practices that refuse the governing operations, step aside, and retreat to the smooth space of commons. The power of capitalism critique in the Arctic arises from refusals and silences, disengagement and withdrawal, tools of resistance at the local level.

6.2 Thinking with, Through and Toward Assemblage

Fig. 6.2 Productive forces

A two-volume work on *Capitalism and Schizophrenia*, consisting of *Anti-Oedipus* (1972/1977) and *Thousand Plateaus* (1980/1987) was an influential success already at the time of publication. In the preface to *Anti-Oedipus*, Michel Foucault (1972/1977: Xii) wrote that the book "can best be read as an 'art', in a sense that is conveyed by the term 'erotic art', for example. Informed by the seemingly abstract notions of multiplicities, flows, arrangements, and connections, the analysis of the relationship of desire to reality and the capitalist "machine" yields answers to concrete questions. Questions that are less concerned with why this or that than with how to proceed". Over ten years later, Brian Massumi (1987), the translator for *Thousand Plateaus*, begins his foreword to the book by writing that "this is a book that speaks of many things" and goes on to suggest that the best way to approach the book is to read it as a challenge.

There is no straightforward way to say what assemblage is and does for you (Fig. 6.2). The problem is not so much that there is a lack of answers, but that there are multiple answers to it. Indeed, Deleuze and Guattari employed a quite deliberate strategy of providing multiple answers to the questions their work raises: the key is not to conform but to create and discover, start from the middle, map connections, processes, and functions. Foucault (1972/1977) writes that the book motivates one to go further.

Given the desire of assemblage thinking to remain an open system and embrace the multiplicity and process of becoming, it feels counterintuitive to try to think about final words to say or conclusions to draw about the Cold Rush and working with critical assemblage analysis. Equally, I recognize the logic of the genre of academic books, and it feels difficult not to write something in order to be able to close this book. I make an attempt to create an ending by drawing on Foucault's and Massumi's framing words, Cold Rush pictures, and my own experience researching Cold Rush as a rhizomatic assemblage.

The work by Deleuze and Guattari indeed speaks of many things and poses several challenges. However, at the same time, it is exactly this capacity to engage with multiplicities that makes their work powerful. When I started researching the Cold Rush in the Finnish Arctic, the rush was palpable, but so too were the contradictions, especially related to a range of values attached to natural resources. I needed a conceptual landscape that would facilitate mapping the evolving and overlapping processes regarding Arctic natural resources and discovering the emergent, ephemeral, and enduring connections and alliances. The concepts of rhizome and assemblage help make sense of this multiplicity without reducing them into a single factor. In addition, they resonated with Foucault's work on discourse, power, and knowledge, as well as ethnography and critical discourse studies.

Critical assemblage analysis, as discussed in this book, is an emergent and incomplete process to examine the multiplicity of ongoing and interconnecting processes that result from the interaction between material things and regimes of signs and meanings. While testing and applying this framework, I noticed that the following three aspects were especially productive in examining the Cold Rush.

Firstly, taking assemblage as a starting point of research means a shift from singularity to multiplicity. This translates into mapping rhizomatic lines, crossovers, and connections. It also means starting from the middle and taking into account all the "stuff" that can be found, both material and symbolic, here and there, now and then. This shift also entails moving an analytical lens from apriory focus to interaction between the "stuff" in a given moment. Sometimes, this means analysis of the historical unfolding of capitalism and colonialism, like the mining assemblage, but at another time, it can mean understanding the nutritional value of a berry or experiencing a long car drive with stops on a cold Arctic night looking for the Auroras. In my previous research, I have often emphasized the analysis of discourses and language use. Shifting to a focus on the interaction between materiality and discursivity opened up a new horizon, where all kinds of stuff, potentially "everything" matters. In practice, this meant, for example, accounting for not only the stories, meanings, and signs but also material objects, embodied practices, affect, spatial organizations, and temporalities. In short, all the elements that the assemblage in question used to function.

However, this was only the start of the mapping. The key question for the analysis is to understand how the particular assemblage functions and what it produces, and to answer these questions requires examining the politics of the assemblage, i.e., the logic or rationality according to which the assemblage is organized and the close analysis of conditions under which the assemblage works. In my case, this translated into a combination of ethnographic, historical, and discourse analysis. All kinds of ways of collecting, analysing, and thinking about the data, relationships, and changes became useful, and to navigate this, I needed to learn new ways of thinking, working, and writing. In my case, fieldnote writing and photographing became key tools to both map but also discover and reflect revaluing processes.

When working with critical assemblage analysis, the key for me is to find a balance between critical and creative, floating, and grounded. When the balance is found, the analysis truly speaks many things about the revaluing of Arctic natural resources and broader ecological, political, and economic processes. And when it does not work, you feel lost, snowed under, and bewildered, which ideally restarts a new round of thinking, time, and energy permitting.

The second key element of assemblage thinking is foregrounding production over representation. This means that the critical question is how an assemblage functions and what it produces, not what it is. In terms of analysis, this means that the focus gears towards mapping processes, interactions, and connections, not on finding out their essences and "truths." This makes ethnography a strong ally in critical assemblage analysis. This was particularly evident for me when examining de/reterritorializing processes related to Arctic natural resources and the various recodings used to revalue them; ethnography helped in seeing how people, projects, companies, and institutions arranged their assemblage with natural resources and among themselves, and what kinds of reconfigurations were needed. Ethnography also helped map the conditions under which the assemblage seemed to work or what obstacles there appeared to be. Here, also the value of discourse analysis was apparent: it shed light on the trajectory and circulation of specific regimes, discourses, and codes, impacting how smooth or stratified the assembling space became.

Finally, the power of assemblage arises from its persistent opposition to dogmatic, tree-type thinking and its equally persistent alignment of critical and creative, rhizomatic thinking. Assemblage allows us to find and examine connections and combinations and to study their relative endurance and ephemerality. Importantly, assemblage invites us to explore the politics of assemblages themselves: the rationales and logics according to which the assemblages are organized to achieve their function.

A fundamental line in Deleuze and Guattari's work is a critique of capitalism and its totalizing expansion. They developed what they called—not unproblematically—"Schizoanalysis" to capture the contradictory ways capitalism is done and undone through flows of desire and to discover the possible resistance and revolutionist lines of flight in specific moments. We could see some fractions of this in the Cold

Rush: the dynamics of de/reterritorialization of natural resources reconfigure the surrounding relations, sometimes opening up new possibilities but also creating instability and uncertainty. The Cold Rush also illustrates how different politics and logics operate through interconnected processes, and the most successful politics are dynamic and adaptive to changing conditions. The "molecular resistance" could be found in the politics rooted in commons, and in people's bodies and embodied experiences. Not everything was convertible for commodity.

The insistence on emergent properties, interactions, and relationships in Deleuze and Guattari's work leads us to engage with ethics that rest on relational ontology (Braidotti 2019). We are products of a multiplicity of forces and are always in the process of becoming. The complex and multiple rhizomes of interdependences of life on this planet have become ever more evident at this moment of climate change. Provisionally, engagement with relational ethics might help us recognize the vitality of the relationships and connections and move beyond and across boundaries, categories, and isolated concepts and contexts. Relational ethics also underlines the specific context of these relationships and helps us to keep in mind the there is no one solution that would function everywhere and for everyone. Rather, it alerts us to reconsider shades, nuances, and connections, especially since relations and by extension ethics, too, are dynamic and evolving. Ethical living, as Deleuze and Guattari (1982/1987) remind us, is about how we navigate and engage with the world we inhabit. Or as Rosi Braidotti (2020) writes: "We are in this together, but we are not the one and the same".

In a spirit of assemblage thinking, there are no final concluding words to deliver. Massuni (1987) suggests that we can ask what new thoughts this text makes it possible to think, and even if the answer for some will be none, that is fine, too. There is always another line to follow and new connections to map. Foucault (1972/1977: xiii–xiv) concludes that Anti-Oedipus is a book of ethics and an introduction to the non-fascist life, and he summarizes seven points for a manual to everyday life. I conclude by citing one of them and by leaving you with a Cold Rush picture (Fig. 6.3):

> Withdraw allegiance from the old categories of the Negative (law, limit, castration, lack, lacuna) which Western thought has so long held sacred as a form of power and an access to reality. Prefer what is positive and multiple, difference over uniformity, flows over unities, mobile arrangements over system. Believe that what is productive is not sedentary but nomadic (Foucault 1972/1977: xiii–xiv).

Fig. 6.3 Interdependencies

References

Braidotti, Rosi. 2019. 'A Theoretical Framework for the Critical Posthumanities'. *Theory, Culture & Society* 36 (6): 31–61. https://doi.org/10.1177/0263276418771486.

Braidotti, Rosi. 2020. "We" Are In This Together, But We Are Not One and the Same. *Bioethical Inquiry* 17: 465–469. https://doi.org/10.1007/s11673-020-10017-8.

Deleuze, Gilles, and Félix Guattari, 1972/1977. *Anti-Oedipus. Capitalism and Schizophrenia*, translated by Robert Hurley, Mark Seem, and Helen R. Lane. London and New York: Continuum.

Deleuze, Gilles, and Félix Guattari. 1980/1987. *A Thousand Plateaus: Capitalism and Schizophrenia*. Translated Massumi, B. London: Athlone.

Foucault, Michel, 1972/1977. Preface. In Gilles Deleuze and Félix Guattari, 1972/1977 *Anti-Oedipus. Capitalism and Schizophrenia* Trans. Robert Hurley, Mark Seem and Helen R. Lane. London and New York: Continuum.

Massuni, Brian 1987. Forewords. In Gilles Deleuze and Félix Guattari. 1980/1987. *A Thousand Plateaus: Capitalism and Schizophrenia*. Translated B. Massumi. London: Athlone.

Index

S. Pietikäinen, *Cold Rush*, Language and Globalization,
https://doi.org/10.1007/978-3-031-63995-1

SPRINGER NATURE

GPSR Compliance

The European Union's (EU) General Product Safety Regulation (GPSR) is a set of rules that requires consumer products to be safe and our obligations to ensure this.

If you have any concerns about our products, you can contact us on ProductSafety@springernature.com

In case Publisher is established outside the EU, the EU authorized representative is:

Springer Nature Customer Service Center GmbH
Europaplatz 3
69115 Heidelberg, Germany

The manufacturer's authorised representative in the EU is Springer Nature Customer Service Centre GmbH, Europaplatz 3, 69115 Heidelberg, Germany. If you have any concerns regarding our products, please contact ProductSafety@springernature.com

Printed and bound by CPI Group (UK) Ltd, Croydon, CR0 4YY
01/10/2026
02225852-0002